NB

Honeybees: A Hive of Information. Annotated Anthology.

ISBN: 978-1-912271-76-4

Published by Northern Bee Books © 2020

Northern Bee Books, Scout Bottom Farm
Mytholmroyd, Hebden Bridge, HX7 5JS (UK)

www.northernbeebooks.co.uk

Tel: 01422 882751

Book design by www.SiPat.co.uk

Honeybees:
A Hive of Information

Annotated Anthology

Graham Kingham

This book is dedicated to my wife and soul mate Cathy and son Daniel, my daughters
Frances, Lindsey and Rachael and their spouses, John, William, Laurence and James.
Grandchildren Enora, Katell, Kara and William who hold the future in their hands.

Contents

Fig 1. Worker bee paused in thought.

Introduction

In the beginning, the virgin queen takes flight and mates with about 15 drones, storing their semen in her *spermathica*. She returns to the colony and starts to lay new eggs around two weeks later. This is the start of a fascinating life story, which unfolds in many fascinating ways. Insects are truly amazing; we could not live without them. I hope you enjoy these articles and they inspire you to inquire further.

These articles have all been published in various beekeeping magazines, mainly local ones. My purpose of writing is to aide me in remembering things, to entertain, educate and on occasions to provoke thought about the plight of honeybees and nature in general. I have tried to look at all things to do with the bee over time, history, and including some biology and botany - mainly pollen identification. Having taken my microscope exam I looked for a project to keep my interest going, together with sharing my findings. Therefore, if you want to know some odd things like what is in bee poo, do bees sleep, what does different pollen look like, what flowers make up honey, why Napoleon chose a bee for one of his symbols, then please read on. I think that we need to become more bee centric and hope that with new evidence regarding our methods and management towards beekeeping, natural selection and changes in how we farm, this will be central to how we all proceed in the future.

Fig 2. A Drone emerging from a cell after 24 days developing,
he has another 15 days before he is sexually mature.

Fig 3. Worker bees just hatched.

Bee bits and pieces

Wing clipping, some thoughts

Many people ask this question and it is one that entomologists talk about frequently.

Humans have *nociceptors*, also called pain receptors found in areas throughout their body that can detect uncomfortable stimulus such as heat, cold and pain, over a certain threshold. This neurological response is called *nociception* and a similar process has been identified in fruit flies however, insects do not have *nociceptors*.

It has always made sense to me that an insect should be able to feel something akin to pain; because it would help them learn to avoid things that could kill them through the unpleasant nature of the experience.

The current definition of pain requires an emotional response. Humans can feel pain without any physical stimulus and are capable of emotions associated with pain; like suffering and terror. Are insects capable of conscious or unconscious experience of emotion? Is consciousness required for emotions? This is where it gets controversial, because how do you quantify if an insect is experiencing an emotion or if insects are conscious?

Insects are hardwired with predetermined behavioural responses to external stimuli, but this is an over simplification.

From the current literature, you cannot definitively say insects have emotion, but there have been many interesting studies on the complexity of insect behaviour. Honeybees, ants and other social insects have complex behaviour and it has long been known that foraging bees will come back to the hive and dance for other bees to express where resources are. One study even shook bees in tubes and measured their agitated responses.

The Cambridge Declaration on Consciousness

We declare the following: "The absence of a *neocortex* does not appear to preclude an organism from experiencing affective states. Convergent evidence indicates that non-human animals have the neuroanatomical, neurochemical, and neurophysiological substrates of conscious states along with the capacity to exhibit intentional behaviours. Consequently, the weight of evidence indicates that humans are not unique in possessing the neurological substrates that generate consciousness. Non-human animals, including all mammals and birds, and many other creatures, including octopuses, also possess these neurological substrates."

So where do insects fall on this scale? Probably further away from *mammals* and even *cephalopods*. However, how do you KNOW? We do not. New studies are coming out and it will be exciting to follow them.

As of right now, the animal testing regulations draw a line between *vertebrates* and *non-vertebrates* (with the exceptions of some *cephalopods*). I do not think any research to date warrants a discussion of moving the line, but at the same time, I do not think we should rule out the possibility that insects are capable of pain, albeit through different *neurological* pathways.

From the article, it admits that nerves are found in the wings, the nerve acts as a pathway for sense messaging to the brain and *ganglions*, although bees might not have pain

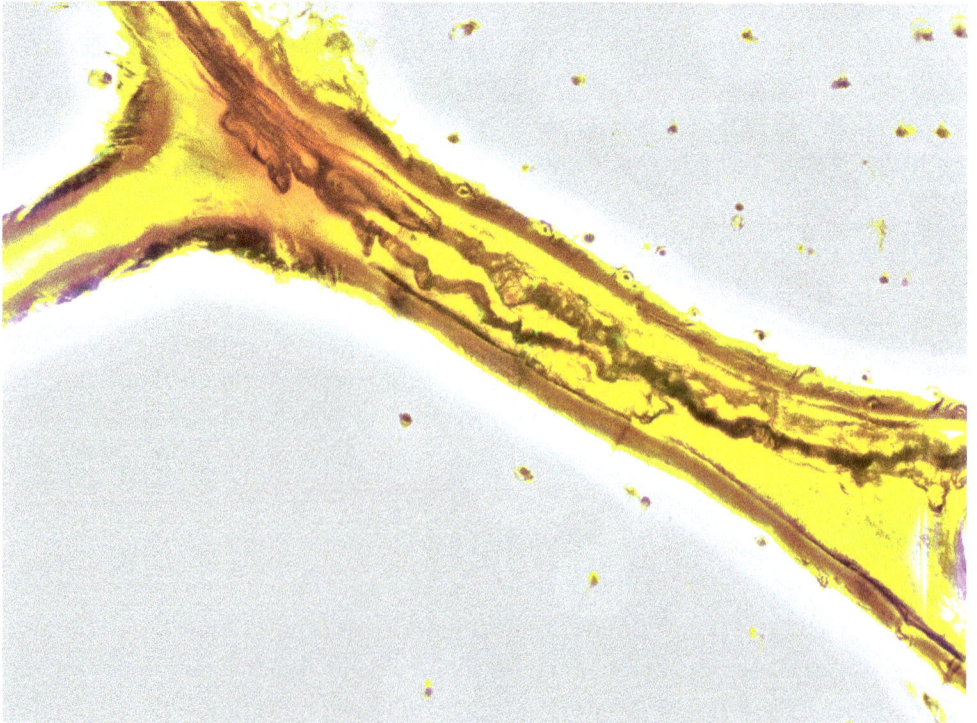

Fig 4. Wing vein, showing tracheoles *and* nerve. X 100 magnification.

sensors they do have lots of sensors, hair and peg like structures to relay messages, such as head movement.

If these are attached to a *neural* pathway they must relay a message, my view for all it is worth is if you cut a nerve on the skin or burn it you get a response, if you break a tooth the underlying nerve is exposed so I think that bees must feel something and I prefer not to clip.

The bee is a social insect and clearly dances, tastes and smells and must feel or observe the buzzing and waggle dance so it has a form of learning and perhaps consciousness dare I say.

I think the major view is they have no consciousness, but more evidence is coming.

We tend to think in human terms and experiences, insects are survivors and specialist in their own environment, and they possibly do have other experiences of the world that we do not. Magnetic tracking, ultra violet light and polarisation detection for example.

We do many things to manage creatures to our own ends, for me clipping is not one

Honey Brewing

Honey has been added to drinks and fermented by its self to form *alcoholic* beverages since time immemorial, the addition to beer can be traced back to 1796, when honey was added at the rate of 385 gms per 23-litre batch, but just what does it contribute to a glass of ale?

For the uninitiated, I will briefly summarise the brewing process.

Barley is first malted via steeping in warm water for a few days, then it is spread out onto flooring to germinate, after a few days the roots and acrospires, (growing tip) are removed and the resulting barley is gentle kilned to dry it out. This process allows the starch that is locked up in the grain to be readily available for the next stage.

The grain is then mashed in hot water for an hour; this is in order for the enzymes in the grain to work to release their sugars into the water. Next, the resulting liquid is drained off into a kettle with added hops, it is then boiled for an hour and a half to extract the lupin in the hops, which give the beer its bitterness and helps to preserve the product. Once it is cooled down, the yeast preparation is added and left to ferment for a week, by that time the beer is racked off into the barrels and left for a further two to three weeks to mature and clear. Then it can be served to the imbiber!

From a technical standpoint, virtually any type of honey can be used in the brewing process. There are over 300 types of honey worldwide, with the colours ranging from water white to dark amber, and the tastes from delectably mild to distinctively bold. Each type of honey contributes something different in terms of end-product colour, aroma, rounding effect and flavour. In lagers, brewers tend to prefer mild honeys such as clover honey. Other floral sources such as alfalfa, wildflowers, sage or citrus are excellent ingredients in porters, stouts and herb or spice beers.

Brewers generally add honey to the kettle toward the end of fermentation to avoid exposing honey to high temperatures for an extended period. This is done to prevent the loss of the honey volatiles, which contribute to the flavour of the final products. Honey beer is often lighter and "crisper" than all-malt beer, but it does not lack character, offering background flavours and aromatic nuances. Honey's carbohydrates are over 95% fermentable and adding honey early in the brewing process will yield a product with no residual sweetness. Honey is often used to obtain a lighter, dryer, more refreshing beer than an all-malt beer.

Through several mechanisms: first, honey contributes its own flavour, second, honey has an impact on how the four basic tastes are perceived and third, honey has a "smoothing" or "rounding" effect on the overall flavour profile. Obviously, the extent to which honey affects the flavour of beers depends upon the type of honey selected (floral source), the amount of honey added and the brewing technique used.

NB. Do not use Eucalyptus honey, as this will produce an undesirable medicinal taste.

Floral Source	Typical Colour	Suggested Use in Beers
Clover	Light Mild	Ales, brown ales, stouts
Alfalfa	Light Mild	Ales, lagers
Sage	Light Mild	Pale ales
Orange Blossom	Light Mild	Ginger, spice beers
Raspberry	White to light Delicate	Ales, spiced / fruit beers
Wildflowers	Medium to dark / Medium to strong	Pale ale, specialty beers
Blended	Medium	Cream stouts, porters

By adding between 3 and 11% will give a subtle honey flavour. At a maximum 30 % a distinctly noticeable honey flavour note will develop. Higher hop ratio, caramelised or roast malts will be needed to balance out the taste at these levels.

Honey contains about 80% fructose, maltose and glucose the remaining amount is made up of proteins, amino acids, vitamins and minerals, giving the flavour compounds.

In honey, wild yeasts and bacteria are ubiquitous, yet they are kept in statis due to the low water content. Averaging 17%, once diluted in wort they are free to proliferate. Boiling honey with the wort destroys these bugs but also hinders the flavour components in the finished ale.

Allow for the increased sugar when formulating your recipe; mash temperature should be higher to encourage dextrin formation, as the sugar content will ferment out leaving a drier, thinner bodied beer. It will also raise the *alcohol* content.

The supermarkets offer a wide choice of 'honeyed ales' on their shelves, why not pick up a bottle and sample one next time you are shopping, or even select one for your loved one as a surprise or reward!

May pollen

Some random samples of pollen were taken from the bottom board of one of my colonies in mid May; there were many yellow lumps in abundance, with a smattering of orange and a pinch of darker brown in evidence. I mixed them all up in some *alcohol* and then made some slides. The yellow staining is from the outer wax coating of the pollen that is visible to us; normally you wick this away and add more clean *alcohol* until you obtain a clear solution. I left this to demonstrate the colouring effect from the wax; it is this, which gives the flower its unique scent. The first slide is pollen mass X 100 magnification, the second is the same but at X 400 magnification, showing some clearer pollen samples. The third slide has been cleaned and stained to show the patterns of the pollen for better identification X 400.

On consulting the pollen books and my reference slide library, I have identified them as, possible Hawthorn or Apple as these look a very similar triangular shape and Buttercup, all of which were in abundance locally. There was a lot of Dandelion about this year; however, I only found two samples in my slide - these are not shown here.

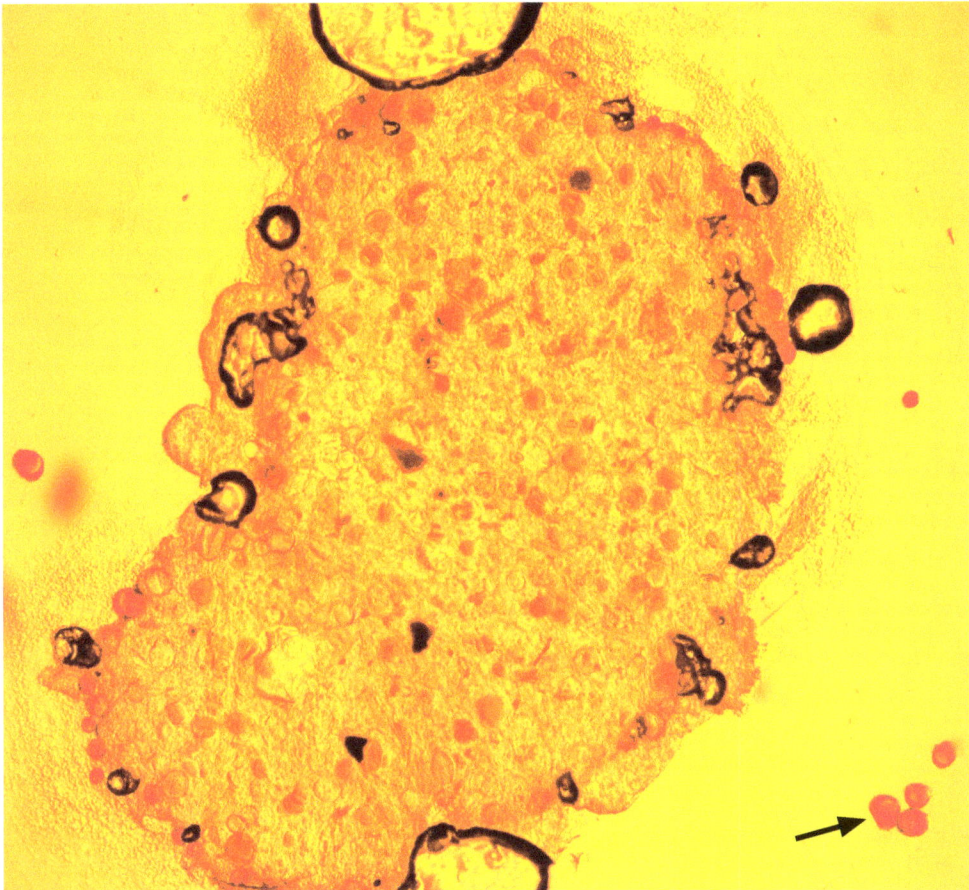

Fig 5. Pollen mass. Stained pollen. X 100 magnification.

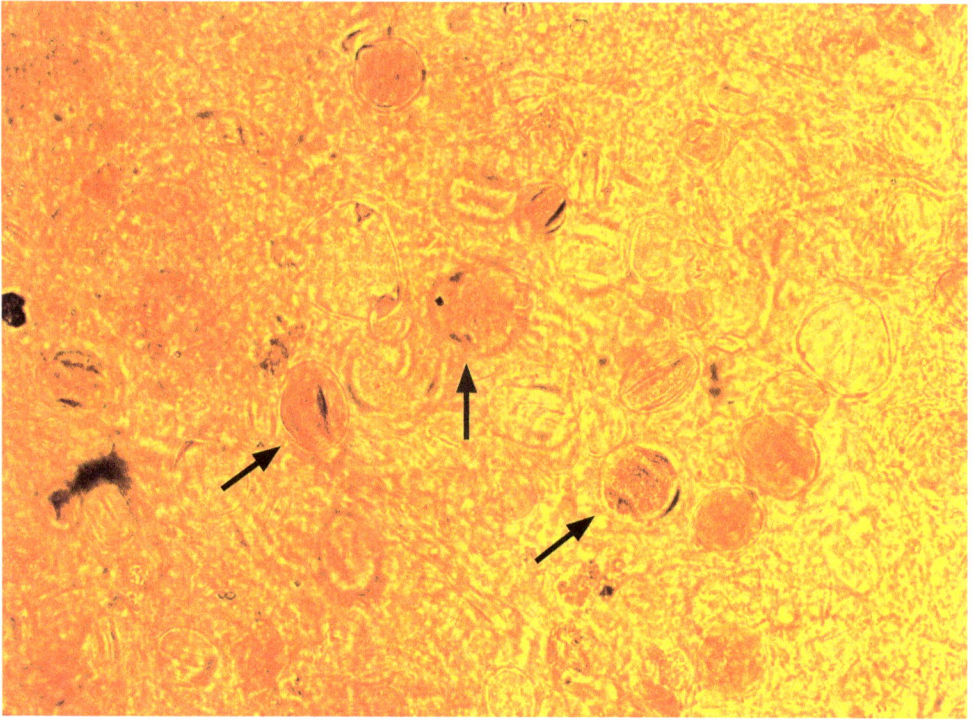

Fig 6. Assorted pollen grains. X 200 magnification.

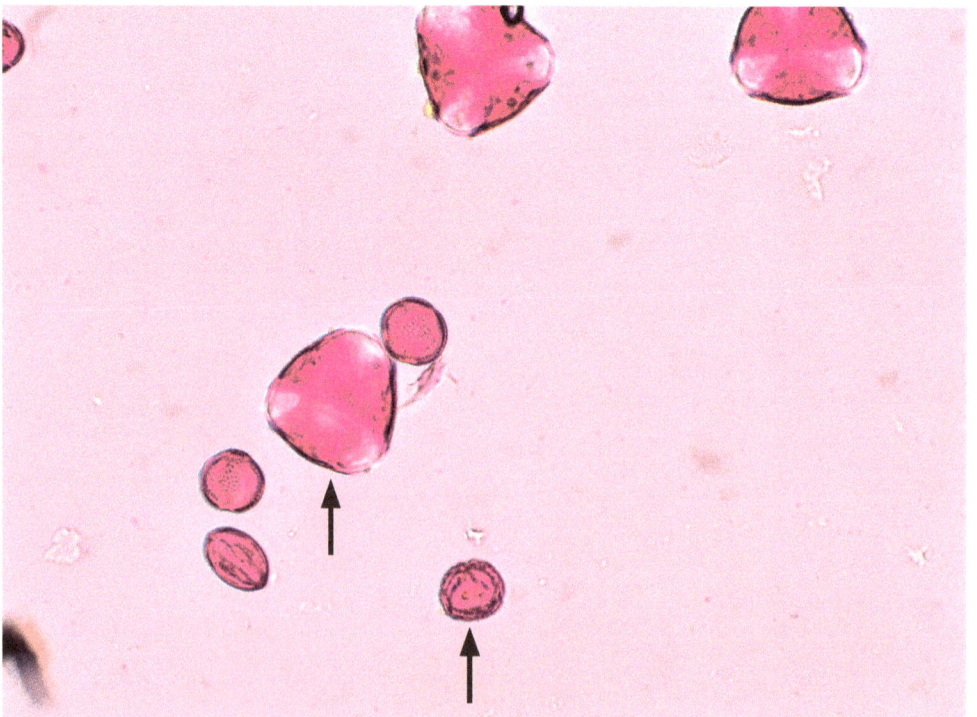

Fig 7. Possible Hawthorn or Apple; Buttercup. X 400 magnification.

Magnetic sense in honeybees

Scientists have recently discovered more about the magnetic sensing in honeybees. They have found iron-based crystals amongst the cells in their abdomen, which work as magneto-receptors. Bee's use the earth's magnetic fields for flight direction and when orientating their comb, in line with the local magnetic field. This evidence shows up when beekeepers have dismantled colonies that have taken up residence in numerous new venues, such as hollow trees or buildings that have to be removed for one reason or another.

Our planet's magnetic field is believed to be generated deep down in the Earth's core.

Right at the heart of the Earth is a solid inner core, two thirds the size of the Moon and composed primarily of iron. At 5,700°C, this iron is as hot as the Sun's surface, but the crushing pressure caused by gravity prevents it from becoming liquid. Surrounding this is the outer core, a 2,000 km thick layer of iron, nickel, and small quantities of other metals. Lower pressure than the inner core means the metal here is fluid.

Differences in temperature, pressure and composition within the outer core cause convection currents in the molten metal as cool, dense matter sinks whilst warm, less dense matter rises. The Coriolis force, resulting from the Earth's spin, also causes swirling whirlpools. This flow of liquid iron generates electric currents, which in turn produce magnetic fields. Charged metals passing through these fields go on to create electric currents of their own, and so the cycle continues. This self-sustaining loop is known as the geodynamo.

The spiralling caused by the Coriolis force means that separate magnetic fields created are roughly aligned in the same direction, their combined effect adding up to produce one vast magnetic field engulfing the whole planet.

Scientists are not sure where all of the bees' magnetic sensors are located. In 2012, an idea emerged that suggests bees may literally "see" lines of magnetism superimposed on their visual image. This does not necessarily mean that the information is collected at the bees' eyes, but it could mean that magnetic and visual data merge in the brain, giving bee navigators a view similar to fighter pilots whose helmets include projections of pathways and flight data. Obviously, there is still a lot to be unravelled here. Researchers have also confused the honeybee by placing strong magnets in their hives, resulting in twisted and messy combs. Similarly, by experimenting with the magnetic field, it appears that bees supplement their navigation with magnetic information. Zoologists

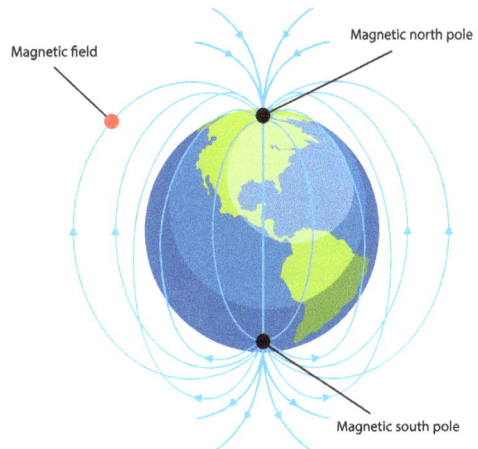

Fig 8. The earth's magnetic field.

disturb the bees' magnetic environment and wait to see if anything interesting happens when scout bees host bee dance parties. It does. In one unusual experiment, tiny magnets were strapped to bees' bellies. These made it hard for bees to follow flight path data they had received back via a waggle dance. Under normal conditions, honeybees can sense variations in magnetism 1/2000 as strong as the Earth's ambient magnetic field. Such fine discernment of just a few nano Teslas (nT) implies (according to Taiwanese researchers Hsu, Ko, Li, Fann, and Lue) that bees may memorize their homestead magnetic field intensity and orientation, allowing them to navigate home from almost anywhere.

There is at least one other place that magnetism may play a role in bee behaviour – drone congregation sites. Before queens and drones mate, the queen is attracted by drone pheromones to meet up in aerial clusters where drones hang out in great numbers. We know that the queen's 'nose' leads her to the mating area. However, the same congregation sites are used year after year by the drones that gather by the thousands but never live more than one season. Repeated use of the same congregation spot – by drones that had never been there before – is baffling. Now it appears that the spot where drones meet, to wait on queens may involve local magnetic anomalies that encourage drones to congregate.

I am beginning to think of the bee as a rather special creature; the worker only lives for 3 weeks outside of the hive, it can see at speed, can find flowers from a distance by smell and taste, gather information by vibration through its feet in the dark hive during the waggle dance.

It uses the ultra violet light and polarizing light together with the magnetic field to map its surrounding routes.

A true specialist in many ways!

Bearding bees

I collected a small swarm, probably a cast with a virgin queen, from a *Buddleia* bush on a hot Sunday afternoon in late June. All went well and as I was able to cut the branch off and place it inside the box, most of the bees stayed in situ. Whilst drinking tea with the host I left the lid ajar for the returning scouts to enter with the aid of the fanning attendants who had raised their rears to the wind.

On arriving home, I put the box in the shade and waited for an hour, yet more tea! The outside temperature was now 26 degrees Centigrade. I then proceeded to gently tip the swarm into the brood box, refit the new frames and closed the roof; all was well; the odd flying bees soon found the new entrance.

Two hours later, I proceeded to check them and found them all bearding on the outside; was this due to the temperature I wondered? I noticed that to attract a swarm you bait the hive with old comb. However, it is recommended that you use clean foundation to house a swarm, to prevent them from filling up the cells with honey that they had ingested from the parent hive leaving no room for development.

Fig 9. Bearding bees. Picture by Catherine Kingham.

Therefore, my thoughts turned to using the only drawn comb I had - some supers. First, I very gently and slowly cupped my hand beneath the bees and pushed them up and over the brood box lip in small amounts, then inserted the drawn supers and closed them up for the night, it was now 9.30 pm.

The good news is that they have stayed put and I will inspect them after a week to see just what they are up to.

The other methods of trying to keep a swarm is to either find the queen and clip her wings; this is not going to work in this case as the queen will need to mate if she is still a virgin. The second method is to place queen excluders underneath the brood box for a few days then remove it; this can work, but the virgin is likely to be a slim bee at this stage, so she still might abscond. If all else fails put it down to the bees knowing best!

A winter's mystery

On Christmas Eve I noticed that my bees were flying at midday and bringing back some white pollen, the day was dry, with slight wind and 10 degrees Centigrade. Come 4 o'clock I noticed a dead bee in the tray that I leave outside of the hive entrance, so that I can observe their behaviour. I had already emptied it at midday.

What plant could this worker have visited? The only plants that I know that still produce pollen in December are *Ivy, Mahonia, Sedum, Heather* and *Viburnam* all of which are in our garden. (We had taken our dog for a local walk and counted 17 plants in flower locally and in the gardens that afternoon). The microscope will have to be consulted.

The samples below show the pollen; most of these grains were not rehydrating, even after 24 hours.

The pollen search was conclusive, after looking at the books and my own pollen collection. The answer was *Heather, Calluna vulgaris*. It has a multisided smooth appearance with some furrows on its surface.

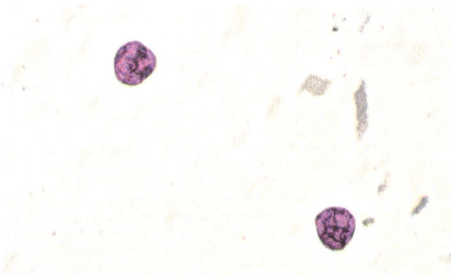

Fig 11. The festive Christmas Eve bee, with white pollen load. X 10 magnification.

Fig 12. Pollen attached by the pollen press on the outside of the bastitarsus X 20 magnification.

Fig 13. Pollen brush on the inside of the bastitarsus showing pollen grains. X 40 magnification.

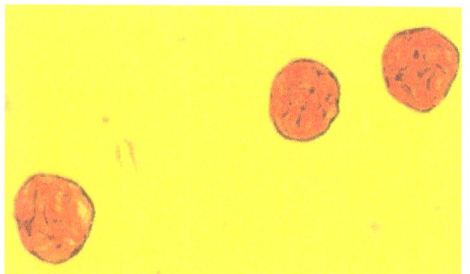

Fig 10. Swollen Heather pollen grain 35 microns. X 400 magnification.

Fig 14. Dried up pollen grains 5-10 microns. X 400 magnification.

Natural Thoughts

The western honeybee, *Apis mellifera*, is one of the most economically important insects, providing essential pollination services for human food security as well as valuable hive products for the apicultural sector.

Recent major losses of managed honeybee colonies at a global scale have resulted in a multitude of research efforts to identify the underlying mechanisms. Numerous factors acting singly and/or in combination have been identified, ranging from pathogens, to nutrition and pesticides. However, the role of apiculture in limiting natural selection has largely been ignored.

As natural selection is the key mechanism of evolution, it will enable any given stock of managed honeybees, irrespective of habitat and genetic background to adapt to each and every stressor as long as the ability to cope with the stressor has a genetic basis so that the respective heritable traits can change in this population over time. Although domestication always interferes by definition with natural selection and apicultural selection has existed for decades, if not centuries, it is argued that beekeeping interference with natural selection in combination with globalization of industrialized apiculture may have now reached levels, where ill effects are inevitable at the colony level.

Specific beekeeping methods, which are likely to interfere with natural selection and possible impact on natural selection, are queen rearing, swarm control, colony management, pest control, mating control, annual requeening, migratory beekeeping, breeding and queen imports. Compared to natural selection, which acts out independently considering the following selections, host-parasite co-evolution, selection for generalists, shifts from vertical to horizontal transmission of genes, trade off scenarios and social adaption.

It is evident that the beekeeper is the most crucial manager of honeybee health. Indeed, beekeepers play the key role in spread as well as diagnosis and control of new and established diseases, for example, treating against ectoparasitic mites, *Varroa destructor*, not only prevents host–parasite co-evolution, but may also add to the exposure to pesticides thereby possibly compromising colony health. In general, the high density of colonies at apiaries promotes disease transmission and impact and the large hives compared to natural nests may have a detrimental impact on colony survival. During routine colony inspections, beekeepers frequently break the natural propolis envelope of colonies, which may compromise social immunity. Apiculture also governs bee nutrition, for example, by placing stationary apiaries in areas with bad forage or by choosing the forage for the bees in migratory beekeeping. The alternation of honey/pollen flows with poor forage periods is indeed a challenge to the colonies to adapt to normal seasonality and may affect resilience to diseases. Replacing diverse honey stores with low-quality sugar water may also affect health and untimely and/or insufficient feeding of honey-depleted colonies for overwintering is an obvious key reason for mortality. Finally, due to the potential role of endosymbionts and the entire associated microbiome of honeybees, treatment of colonies with acaricides, antibiotics, and even sugar feeding may interfere with natural population dynamics of such associated prokaryotes.

While treatment against disease is helpful, it nevertheless prevents natural selection for improved host resistance and tolerance. In particular, the common practices of removing drone brood to control *Varroa destructor*, castrates colonies, thereby preventing that well-adapted ones spread their genes in the population. This seems significant because recent evidence suggests substantial local adaptations of honeybees enhancing colony survival and reducing pathogen loads.

These genotype–environment interactions, including immuno-priming of eggs by the queen in response to pathogens in the hive, are routinely and constantly disrupted when queens or colonies are moved over large distances, for example, from Southern Italy to Finland, as part of international apicultural trade. Indeed, the industrial production of tens of thousands of queens annually, which are nowadays exported at a continental and even global scale, clearly interferes with any local adaptations. Therefore, 'think globally, but breed locally' appears an adequate suggestion for honeybee breeders to take advantage of natural selection and to foster local adaptations.

In artificial insemination, breeders choose drones of the right age, which obviously have not made it yet to drone congregation areas and may thus not have the full reproductive potential. At isolated mating apiaries, only a few drone-producing colonies are provided, which are often headed by sister queens, thereby clearly limiting the full potential of the highly polyandrous mating system of honeybees to generate subfamilies with ample genotypic diversity and respective derived benefits. The equal number of mating's of wild and managed queens suggests that the system has evolved to provide optimal genetic variation of colonies, but will fail to deliver with closer genetic similarity of the drones and reduced mate numbers.

The build-up of a stable host–parasite relationship is strongly favoured by vertical transmission of the parasite and is unlikely to occur when horizontal transmission is the predominant route. Indeed, shifts from vertical to horizontal transmission are known to increase pathogen virulence. However, the common practice in commercial beekeeping in most countries to routinely requeen colonies annually or every two years limits the full adaptive potential of vertical transmission. After requeening, parasites are confronted not only with an entirely new queen genotype, but also with novel genotypes of the drones that the queens have mated with, assuming natural queen mating at apiaries and unrelated drone/queen sources. This may have caused shifts from vertical to horizontal transmission with respective consequences for the virulence of honeybee parasites.

Commercial breeders select against swarming, defensive behavior, and propolis usage, thereby probably compromising colony defence and social immunity. Indeed, in Africa, where the majority of honeybee colonies are not kept by man and where beekeepers are mostly side users not interfering with natural swarming, queen rearing etc., the virtually nonbred local subspecies have less desirable beekeeping traits, but a superior health compared to European ones. This supports the notion of a trade-off scenario between commercially desired traits and bee health. In particular, queen failure is one of the foremost mentioned causes of honeybee losses and may be linked to breeding, because queen breeders usually ignore choices made by colonies and choose *larvae* based on right age alone. The natural reproductive cycle of a colony, including hormonal and nutritional aspects, determines timing and development of drones and new queens and often lays outside of the time window for commercial queen rearing.

Moreover, during emergency queen rearing, the choice of the bees is not at random; instead, subfamilies, which are rare in the work force, are significantly more likely to end up as queens. As such royal subfamilies are rare, human choice of *larvae* based on appropriate age alone is likely to miss those and instead offers only suboptimal choices for the bees.

Fig 15. The perfect bee? Probably a local hybrid that has adapted to its environment. Just as nature intended.

Moreover, breeding for *Varroa destructor*-resistance over twenty years has still not resulted in survival of untreated colonies, but natural selection has delivered multiple times, thereby suggesting that breeders should choose traits favoured by natural selection. This suggests fundamental conceptual flaws in both commercial honeybee queen rearing and breeding. As the fitness of a honeybee colony clearly is the number of surviving swarms as well as the number of successfully mating drones, all other traits are only tokens of fitness, the selection by beekeepers for low swarming tendency of colonies and removal of drone brood, mainly to combat mites *Varroa destructor*, remain probably the key factors in limiting natural selection.

Conclusions

It is obvious that taking into account natural selection will not solve all of the various problems for apiculture, but instead it is considered a main issue in itself now. As natural selection is the differential survival and reproduction of individuals due to differences in phenotype, future efforts to enhance managed honeybee health should take into account the central role of apiculture in limiting natural selection and compromising colony health via adjusted keeping and breeding of local bees. Here lies a great opportunity for beekeeping in several countries, where economic constraints are no longer leading, as beekeeping has become a hobby sector, with dispersed and small apiaries being the rule. Sustainable solutions for the apicultural sector can only be achieved by taking advantage of natural selection and not by attempting to limit it.

Summaries of a paper written by Peter Neumann from Basle University, who has kindly agreed to its publication. *https://doi.org/10.1111/eva.12448*

An alternative slant on honeybee history

I have read an excellent book about Josephine Bonaparte by Kate Williams, Josephine: Desire, Ambition, Napoleon, in which she records Napoleon's use of the honeybee as a symbolic emblem for his reign as emperor.

Once he decided that the icon of his reign would be the eagle of the Caesars, the bird of power and victory. Hunting for another symbol, something suitably memorable to outdo the fleur-de-lys, he chose the bee to evoke Childeric, the fifth-century king of the Franks. When a mason had found Childeric's tomb in 1653, it was filled with precious objects and more than three hundred gold bees. (Now believed to be Cicadas) Napoleon felt that the bee was a symbol of resurrection, immortality and royal authority. This became another ideal icon for his reign. Fabric and carpet makers were at

Fig 16. Gold bees on the red surround. This file is licensed under the Creative Commons Attribution-Share Alike 3.0 Unported, 2.5 Generic, 2.0 Generic and 1.0 Generic license. https://commons.wikimedia.org/wiki/User:Katepanomegas.

once set to weaving bees into every piece of material that would take them. He commissioned Fontaine to design an imperial state coach covered with stars, laurel leaves, and bees and bearing an eagle and Charlemagne's crown at the top. Embroidered bees, gilt bees, bronze bees, all hovered, and buzzed on the curtains, floor coverings, wall hangings, books, and furniture of the Tuileries.

Several artists spent time studying the fluerons, which were part of the Childeric tomb treasure. The treasure was still housed in the Bibliothèque Nationale, where it had remained since the Austrian Emperor had given it to Louis XIV in 1665. For some purposes, the original shape of the "bees" was acceptable, but not for the imperial cloaks which the Emperor and Empress would wear during the coronation ceremonies. Jean-Baptiste Isabey, best known as a miniature painter, and a close friend of the Bonaparte's, was charged with the design of these very important garments. He was also responsible for the design of the garments to be worn by the highest-ranking dignitaries who would attend the ceremony. Isabey found that the Childeric "bee" was too compact and too lacking in detail to give the desired effect when embroidered in an all-over pattern in gilt thread on the red velvet of the coronation cloaks. Isabey developed a new bee design in a larger size, which is, seen from the top with partially open wings. According to legend, the bee never sleeps so it has also come to imply vigilance and zeal – both attributes Napoleon was happy to own.

Fig 17. A closer view of Napoleons coronation painting by Jacques Louis David. The red robes showing the embroidered bee emblems. Wikipedia public domain.

Honey magic

Honey has been in use as a medicine both internally and externally for over 2000 years; however due to the lack of patenting of the products, there has been little money available for research, and this is now starting to change. This article summarises some of the known facts to date about its magical properties.

During the past dozen years or so, researchers have been discovering how honey heals and have identified more than 250 clinical strains of bacteria honey can kill. These include *Escherichia coli* and *Helicobacter pylori*. Perhaps most impressive is the finding that most raw honeys can effectively kill MRSA, the much-feared "superbug". Honey either kills by bactericidal activity or inhibits the growth of a wide range of bacterial and fungal species (bacteriostatic). Honey has four main properties that help it to exterminate bacteria: osmosis, high acidity, hydrogen peroxide activity, and a variety of phytochemicals that are derived from both the honeybees and the plants they visit.

The Power of Osmosis

Honey works differently from antibiotics, which attack the bacteria's cell wall or inhibit intracellular metabolic pathways. Honey is hygroscopic, meaning it draws moisture out of the environment and thus dehydrates bacteria. This happens because honey is a saturated (or supersaturated) solution of sugars, with 84 percent of the honey being a mixture of fructose and glucose. The water content in honey is usually only 15–21 percent by weight. The strong interaction of these sugar molecules with water molecules leaves very few of the water molecules available to support the survival—let alone growth—of microorganisms.

Low pH for Health

The standard measure of the acidity or alkalinity of a solution is expressed as pH (potential of hydrogen). Aqueous solutions at 25°C with a pH less than7 are considered acidic, whereas those with a pH greater than 7 are considered alkaline. Honey is very acidic. Its pH is between 3 and 4, which is roughly the same pH as orange juice. Most types of bacteria thrive at pH levels between 7.2 and 7.4 and cannot survive at levels below pH

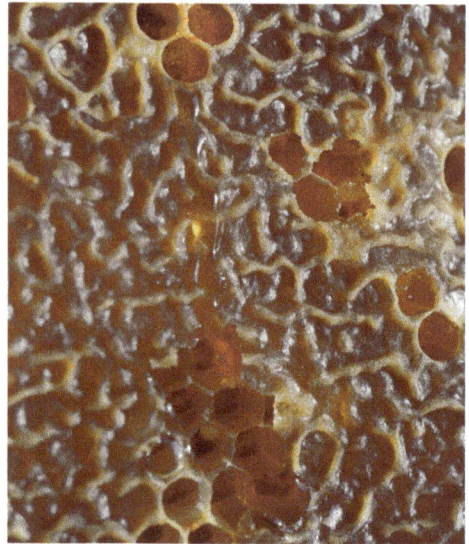

Fig 18. Sealed honeycomb.

4.0, like those found in honey. However, if the honey is diluted (for example, by the release of body fluids from a wound), it may become less acidic, allowing bacteria to grow again. The low water activity of honey could be expected to dry out a wound, but this is not the case. The osmotic power of honey described above draws out fluid from the plasma or lymph in the tissues that underlie the wound. This activates the enzyme glucose oxidase, which enables the honey to produce hydrogen peroxide, an important factor in inhibiting bacterial growth and promoting immune activity.

Hydrogen Peroxide

Hydrogen peroxide is created in the atmosphere when ultraviolet light strikes oxygen in the presence of moisture. Water (H_2O) combines with the extra atom of oxygen to be become hydrogen peroxide (H_2O_2).

Aside from being known as a powerful oxygenator and oxidiser, a special quality of hydrogen peroxide is its ability to readily decompose into water and oxygen. Hydrogen peroxide reacts easily with other substances and is able to kill bacteria, fungi, parasites, viruses, and even some types of tumour cells.

It also has an extraordinary capacity to stimulate oxidative enzymes. Oxidative enzymes can change the chemical component of other substances (like viruses and bacteria) without being changed themselves.

H_2O_2 and Honey

Hydrogen peroxide is made naturally in honey by an enzyme called glucose oxidase, which is added to the plant nectar by the bee. Glucose oxidase is secreted from the hypopharyngeal gland of the bee into the nectar to help formulate honey from the nectar. It is believed that the hydrogen peroxide is used as a sterilizing agent during the honey's ripening process. It is interesting to note that full-strength honey has a negligible level of hydrogen peroxide, because this substance is short-lived in the presence of the transition metal ions and ascorbic acid in honey, which cause the hydrogen peroxide to decompose to oxygen and water. Glucose oxidase has been found to be practically inactive in full-strength honey, giving rise to hydrogen peroxide only when the honey is diluted. When it comes to clearing infections, honey supplies low levels of hydrogen peroxide to wounds continuously over time as opposed to a large amount at the moment of treatment. In essence, it becomes a powerful yet effective "slow-release" antiseptic at a level that is antibacterial but does not damage tissue.

Varying Antibacterial Potential

There are differences in the antibacterial activity of different honeys. A method was devised to determine the inhibine number of honeys (a measure of their antibacterial activity). The inhibine number is the degree of dilution to which a honey will retain its antibacterial activity, representing sequential dilutions of honey in steps of 5 percent ranging from 5 to 25 percent. For example, one variety of honey may be able to kill a specific type of bacteria at a 15 percent dilution but not at 20 percent. Another honey can be diluted to 25 percent and still be effective at killing that same bacteria. The second honey would have a higher inhibine number. Studies measuring the inhibine number of honeys have shown that antibacterial activity can vary considerably: up to a hundredfold. Some honeys are no more antibacterial than table sugar, while others can be strongly diluted and will be able to kill or inhibit the growth of bacteria.

Phytochemical Factors

While important, hydrogen peroxide activity does not account for all of honey's antibacterial activity. Honey also contains phytochemical factors, which are chemical compounds such as a carotenoid that occur naturally in plants. They are found in the nectar that the bees collect. Not only does each plant species supply specific phytochemicals, but the chemical activity can also vary from plant to plant.

Which Honeys Have More Antibacterial Activity?

Honeydew honey from the conifer forests of the mountainous regions of Germany and central Europe has been found to have particularly high antibacterial activity. In a study on the antibacterial activity of honey on *Helicobacter pylori*, the researchers found that Black Forest honey scored the highest among eight honeys in antibacterial activity. Darker honeys (such as buckwheat, clover, manuka, ling heather, and *borage*) tend to have higher antibacterial activity than paler honeys like acacia, and sage.

The Power of Manuka

Dr Peter C. Molan of the Department of Biological Sciences at the University of Waikato in New Zealand discovered that honey made from the flowers of New Zealand's manuka trees (*Leptospermum scoparium*) seems to be especially powerful at killing bacteria and does not depend on hydrogen peroxide activity to do so). Although other honeys may well contain specific antibacterial properties, Dr. Molan believes that

Cocci

Others

coccus   diplococci   diplococci encapsulated Pneumococcus   Staphylococci

enlarged rod Fusobacterium

streptococci   sarcina   tetrad

Vibrio   Comma form Bdellovibrio

Bacilli

coccobacillus.   bacillus

Club Rod Corynebacteriaceae   Helical form Helicobacter pylori

diplobacilli   palisades.

Corkscrew form Borrelia burgdorferi

Streptobacilli

Budding and appendaged bacteria

hypha   stalk

Filamentous   spirochete

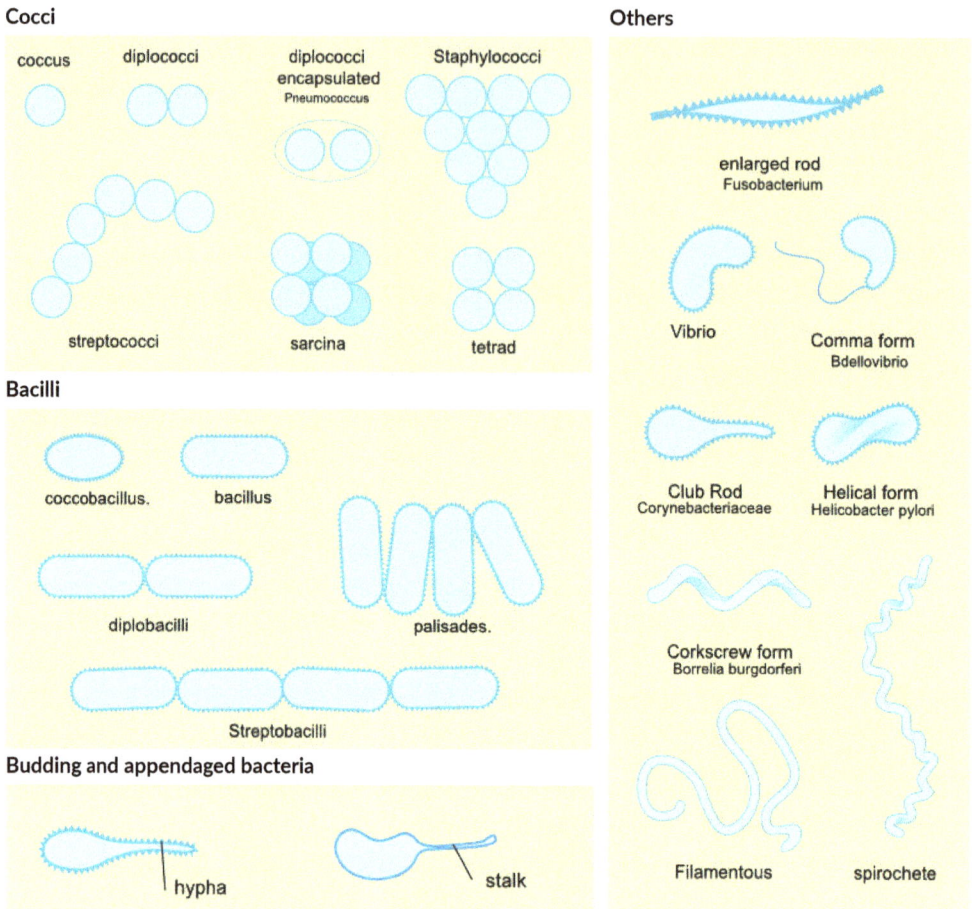

Fig 19. Bacteria types. Wikipedia https://creativecommons.org/ licenses/by-sa/3.0/ Public domain by LadyofHats.

the antibacterial component found in *Leptospermum* plants is unique. That is why he decided to call it the Unique Manuka Factor (UMF) and developed a test to evaluate the UMF levels reflecting antibacterial strength as distinct from hydrogen peroxide activity. Manuka is an evergreen shrub growing up to 3m by 3m it does not like very dry soil, likes shaded roots and full sun on the foliage, and tolerates severe coastal conditions. Hardy to minus 23°C. The Tregothnan Estate in Cornwall has been breeding the rare Manuka plant since the 1880s; the plants are freely available in the UK. Myrtaceae *Leptospermum scoparium* is a member of the myrtle family, which include myrtle, clove, guava, acca, allspice, and eucalyptus, all very aromatic plants.

The Properties of Manuka

UMF-rated Manuka honey has several outstanding healing properties. In laboratory tests, UMF-rated Manuka honey inhibits the growth of *Helicobacter pylori*, a species of bacteria that is linked to stomach ulcers. It also kills *Citrobacter freundii*, *Escherichia coli*, *Proteus mirabilis*, and *Streptococcus faecalis*. The peroxide activity of other honeys (including ordinary manuka varieties) is not effective against these bacteria, which led Dr. Molan to conclude that manuka honey was superior to other honeys in treating infected wounds. Although both hydrogen peroxide and UMF-rated manuka honey can totally inhibit the growth of *Staphylococcus aureus* during an 8-hour incubation period, manuka honey has been found to work twice as fast as the other honey varieties. UMF-rated Manuka honey is also significantly more effective than other honeys against *Streptococcus pyogenes*, which causes sore throats.

The Testing Criteria

Manuka honey samples undergo two special tests developed by the Honey Research Unit at the University of Waikato before they can be classified as

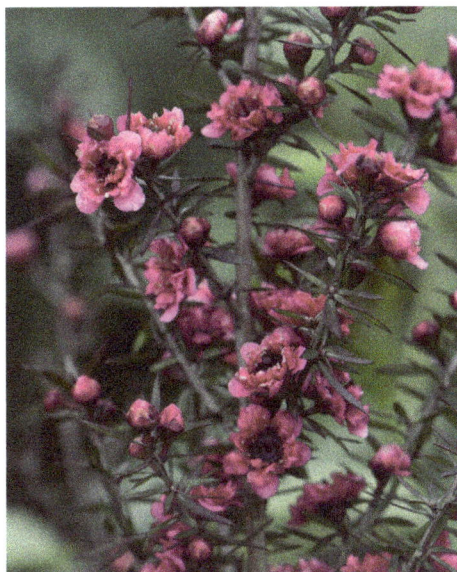

Fig 20. Manuka flower by kind permission of RHS © RHS.

UMF-grade honey. The first test is designed to reflect "Total Activity." A Total Activity rating shows all the antibacterial activity of the honey without distinguishing between enzyme activities (that produces hydrogen peroxide) or UMF activity. Any type of honey can have a Total Activity rating. The second test is designed to reveal specific UMF activity in honey. After the Total Activity rating is established, catalase is added to the honey sample to destroy any enzyme activity. Then the UMF activity of the honey is measured, which can range from a rating of 0 to 20. A rating of 0 to 4 shows that the UMF factor is "not detectable," whereas a rating from 5 to 10 shows that the honey can be used for "maintenance levels" only. Manuka honeys with UMF ratings of 10 to 15 are considered "useful" for most types of wound treatment, while those of 15 and above are classified as "superior" by the Honey Research Unit. Manuka honey with a rating of 10 or above is considered adequate for clinical use. Only licensed companies that meet specific criteria delineated by the Honey Research Unit and the AMHA can use both the name and trademark.

Medical Grade Manuka

In addition to the UMF rating, some manuka honey is classified as "medical grade" honey. Sold under brands like Actilite, Advancis medical UK. It is also the raw material for government-approved sterile dressings made up of active manuka honey and alginate fibre like the Apinate brand produced by Brightwake Ltd. Sterilization of therapeutic honeys is achieved by gamma irradiation. Unlike the heat treatment used on supermarket honeys, which destroys the enzyme responsible for producing hydrogen peroxide. The processed honey is then shipped to a licensed manufacturer that produces honey-based products like antiseptic creams and wound dressings.

Doctors have not only found that honey rapidly clears infections from wounds, but it also actually promotes healing. Laboratory studies have found that honey inhibits the growth of more than sixty types of bacteria, including those most commonly associated with wounds. It also works where bacteria-resistant antibiotics do not, such as killing the dreaded *Streptococcus pyogenes*, or "flesh-eating" bacteria. They have also found that honey rarely produces adverse side effects and is far more cost effective than traditional medicines used to treat both minor and serious wounds. Why is honey such a miracle wound healer? Laboratory and clinical research offer the following reasons: bacteria cannot live in the presence of honey. The osmotic pressure (pressure that sucks a solvent through a membrane of a cell into a denser solution) that honey naturally exerts, removes water molecules from bacteria, making them shrivel up and die. Honey placed on a wound creates a physical barrier through which bacteria cannot pass. When honey is diluted with water, the glucose oxidase it contains becomes active and produces hydrogen peroxide, a powerful antibacterial agent. The sticky texture of honey prevents dried blood from adhering to the bandage. Dressings can be removed from the wound without hurting new skin cells. Adverse side effects of honey-based wound dressings are extremely rare.

Deodorization

One of the most psychologically debilitating aspects of an infected wound is the strong odour. This is caused by malodorous substances produced by bacteria, including ammonia, sulphur compounds, and certain types of amino acids. On one level, honey kills the anaerobic bacteria (bacteria that thrive in a low-oxygen environment) found in chronic wounds. In addition, honey contains copious amounts of glucose, a substrate (a substance acted upon and changed by an enzyme) metabolized by bacteria in preference to amino acids. This is why honey can deodorize wounds so rapidly and effectively. In some cases where wounds grow exuberantly like a fungus, honey was found to be the only way to control the offensive odour.

Wound Debridement

Like other moist wound dressings, honey facilitates the debridement—or the removal of dead tissue and foreign matter—from a wound. It accomplishes this feat through the process of autolysis, or the digestion of cells by different types of enzymes. This osmotic action also provides a constantly replenished supply of proteases (also known as proteolytic enzymes) that also aids in wound debridement. In addition, osmosis prevents the new tissue from becoming too soft due to an accumulation of moisture. When honey is placed on a wound, osmosis creates a layer of fluid beneath the bandage made up of the honey diluted in plasma or lymph. This makes it impossible for the bandage to adhere to the wound. When the dressing is removed, new-growth tissue is not torn away.

Honey: The Downsides

Although honey has numerous advantages over other treatments in wound healing, it is important to mention that it has several disadvantages as well. Honey can become more fluid at higher temperatures. It may liquefy at wound temperature, thus making it less effective. The risk of liquefaction can cause leakage and create a terrible mess. Preparation of homemade honey dressings is not easy. A small number of patients (approximately 5 percent) have experienced a temporary stinging sensation in the wound, which is a major cause of discomfort. The availability of commercially produced sterile dressings impregnated with honey (such as Medihoney prepared gels and dressings, Activon Tulle, and Comvita's Apinate dressing) enables doctors to avoid many of these problems. Some of these new dressings contain a hard gel that swells to a softer gel as it absorbs liquid from the wound. This enables the honey to keep in contact with the wound itself, while preventing the moisture from leaving the bandaged area.

If you wish to use honey for any medical treatments please consult your doctor first, you never know they might just be beekeepers as well.

My thanks go to his widow, Alyson Molan for her kind permission in allowing me to use some of his material.

Insects abound - the order of things!

An estimate of insect abundance is not constant. One reasonable estimate places the total number of insects at about one quintillion (1,000,000,000,000,000,000), equal to about 2.7 billion tons of insects, outweighing the human population by nearly ten times. Biologically speaking, there are a lot of insects. No other animal group is even within a magnitude of scale to the insects in this regard. Thus, insects can be argued to be the most successful group of animals, based on outnumbering all competitors.

By contrast, if we simply say that the number of insect species is approximately one million, then insects would represent more than 75% of all animal life on the planet. So if the truer number is closer to five million that means insects is even more dominant, approaching 90%–95% of the existing animal life on the earth, living in every conceivable environment from volcanoes to glaciers. Insects help us by pollinating our food crops, decomposing organic matter, providing researchers with clues to a cancer cure, and even solving crimes. They can also harm us by spreading diseases and damaging plants and structures.

The class Insecta encompasses all of the insects on the earth. It is most often divided into 29 orders.

The genus and species names are always italicised, and used together to give the scientific name of the individual species. An insect species may occur in many regions, and may have different common names in other languages and cultures. The scientific name is a standard name that is used by entomologists around the world. This system of using two names (genus and species) is called binomial nomenclature. *Taxonomy* is a hierarchical system for classifying and identifying organisms. This system was developed by Swedish scientist Carolus Linnaeus in the 18th century and now updated to include RNA structure.

Linnaeus's *taxonomy* system has two main features that contribute to its ease of use in naming and grouping organisms. The first is the use of binomial nomenclature.

Fig 21. Capture drone.

This means that an organism's scientific name is comprised of a combination of two terms. These terms are the genus name and the species or epithet. Both of these terms are italicised and the genus name is capitalised. For example, the scientific name for humans is *Homo sapiens*. The genus name is *Homo* and the species is *sapiens*. These terms are unique and no other species can have this same name.

The second feature of Linnaeus's *taxon*omy system that simplifies organism classification is the ordering of species into broad categories. The major categories include Kingdom, Phylum, Class, Order, Family, Genus, and Species. These categories have been updated to include Domain in the *taxon*omic hierarchy. Domain is the broadest category and organisms are grouped primarily according to differences in ribosomal RNA structure. *Taxon*omic categories can be further divided into intermediate categories such as subphyla, suborders, superfamilies, and super classes. An example of this *taxon*omy scheme is below. It includes the eight main categories along with subcategories and super categories. The super kingdom rank is the same as the Domain rank. Honeybee occupations break down into one of three categories: worker bees, drones and the Queen bee.

Classification: The meaning of *Apis mellifera* in English is honey bearing. Their phylogenetic tree (evolutionary) is based on morphological concept (due to form or structure)

Domain: Eukaryota - Honeybees are members of the domain Eukaryota because their cells have a membrane-bound nucleus.

Kingdom: Animalia - Honeybees are members of the Animalia because they are, multicellular, heterotrophic and motile at some point in life.

Phylum: Arthropoda - Honeybees are members of the Arthropoda because they have segmented bodies and a jointed exoskeleton.

Epiclass: Hexapoda - Honeybees are members of the Hexapoda because they have six walking appendages and three body segments.

Class: Insecta - Honeybees are members of the Insecta because they do not have any muscles past the first segment of their antenna; they have something called Johnston's organ; and they have an internal structure inside the head called a tentorium.

Order: *Hymenoptera* - Honeybees are members of this group because they have "membranous wings and their abdomen's first segment is fused with the last segment of the thorax.

Family: Apidae - Honeybees belong here because they have a *corbicula* (pollen basket) on the outside of each hind appendage.

Genus: *Apis* - Latin for bee. All members of this genus are actually honeybees

Species: *mellifera* - Latin for Honey bearing.

Subspecies: A. *m. carnica* (Carniolan)

Subspecies: A. *m. caucasica* (Caucasian)

Subspecies: A. *m. ligustica* (Italian)

Subspecies: A. *m. mellifera* (European dark)

It is all in the name

The Roman statesman and scholar, Marcus Terentus Varro who served as Julius Caesar's librarian drafted a theory known as the 'Honeycomb Conjecture'. Being a beekeeper he noted the hexagonal shape of the comb and proposed they built them that way for the sake of efficiency. No other inter-locking shape would hold so much honey with so little wax. It was not until 1999 that a mathematician proved him right. *Taxonomists* came up with the genus '*Varroa*' when naming the dreaded mite, forever associating the old Roman with the deadly threat to the bees he so admired.

Fig 22. Varroa without its Toga! X 30 magnification.

Not quite the right buzz

Nature has worked closely with the bee species during the evolution of plants. In exchange for nectar, pollen is transported on the bee for cross fertilisation of neighbouring plants of the same type. One plant family, the Solanaceae, that includes tomato, nightshade, aubergine and blueberry, have taken this process a step further, making the buff tailed bumblebee its preferred partner. Due to the way the flowers are designed it requires the right frequency of buzz to shake off the majority of pollen on to the bees back hairs and only bumble bees are equipped to do so. They hold onto the protruding petals and bite the flower, leaving behind the tell tale sign of brown marks, they then buzz their wings for a few seconds only; this causes the pollen to be vibrated off. However, it is not the same frequency used as when flying. The flowers have what are called 'poricidal *anthers*', a design that holds their pollen in tiny chambers accessed only by a small pore at one end. While some shakes out naturally over time, allowing for some self-fertilisation it requires the right vibration to resonate and release a spray from the pore. The final twist is that tomatoes do not produce any nectar. The honeybee just does not cut the mustard (tomatoes) here! Take a look at your greenhouse tomatoes this year. If all of the 10 flowers on a stem have been pollinated you will get a bumper crop; however, like most of us you will more than likely get 3-4 only out of the potential 10; the solution is to leave the doors of the greenhouse open and encourage the humble bumbles in, or alternatively, shake you plants often!

The A Bee C

When entomologists study problems in insects they tend to look at the 4P's first. Parasites. Poor nutrition. Pesticides and Pathogens. Most insects tend to harm the plants by eating them; however bees are unique by working alongside the flowers, only eating nectar, which does no harm and aids in pollination. This symbiotic relationship rewards both parties. The down side of this is that systemic treatments to kill off other bugs are found in the pollen and nectar and are potentially dangerous to the bee population. A few other letters have also found their way into the equation. A stands for the new UK threat, Asian hornet. N for nesting habitat, so easily lost to domestic farming practices. I for Invasive species, think of *varroa* and the Asian hornet. C stands for Climate change; we had an extended winter in 2018, the longest in living memory. So what do these letters spell out? PANIC! As good an anagram as any I suppose. Please feel free to fill in your own list within each category.

These articles have been inspired by the book *BUZZ: The Nature and Necessity of Bees* by Thor Hanson.

Entomology, etymology, the honeybee and words

The origin of words and their meaning can often be found in the mists of time. The Welsh language has many bee related words that are found in early history; part of the reason for this being that property was passed down in the family and therefore it had a value; in order to describe it, a word was needed. In fact, early records are available in law courts relating to disputes over property, including bees and skeps.

The English term *honie bee* was used as early as 1566 according to the Oxford English Dictionary.

'Some words that describe honey: milit, mez, mesi, mit, mitsu, mi, mil, miele and mel, all share a single root across Indo-European languages, tracing the history of the western honeybee, *Apis Mellifera*, wherever our fore fathers took them. At some point the Germanic language describes honey by its colour; its origin comes from the Old Norse hunung, then the Old German became honing, the Old English word was, hunig, which is how we arrived at today's English version, honey.'

(The above paragraph, which inspired this article, was taken from 'A Honeybee Heart Has Five Openings', by Helen Jukes. With kind permission. This book is a fascinating mix of biography, autobiography and natural history, entwined with a meditation on what it means to keep, to love and to find one's place in the world. At its core is the author's experience of learning the art of beekeeping - and while other writers create magic with majestic landscapes or charismatic creatures, this book creates a sense of heart-felt wonder with only a hive full of insects and a suburban back garden. A pleasure to read, not aimed at educating the beekeeper but a journey and de-stressing adventure from life and work through the wonderful world of bees.)

A list of some of the common used words and phrases & their meanings, with bees in mind:

Bee: 'Bee in your bonnet' - being preoccupied. 'Busy bee' - a person who is industrious. 'Float like a butterfly, sting like a bee' - to be deceptive in appearance. 'The bee's knees' - are the best. 'None of your bee's wax' -none of your business. 'Queen Bee' - a woman in a position of dominance. 'The birds and the bees' -euphemistic for sex education. 'Beeline' - going in a direct line. 'Beeswings' - a crust of tartar formed on old port.

Honey: 'The land of milk and honey' - a promise of plenty. 'Honey-tongued or honeyed' - smooth, persuasive and seductive. 'You can catch more flies with honey than with vinegar' - having a sweetness of nature is a better proposition. 'Honeymoon' - signifying a period of a month after marriage.

Buzz: 'Buzz word' - a vogue word. 'Buzz off' - to go away. 'Get a buzz out of something' - to get excited.

Hive: 'A hive of activity' - shows signs of great activity. 'Hive off' - to be transferred from a larger group.

Sting: 'Sting in the tail' - an unexpected and unpleasant end. 'Sting into action' - being goaded or to incite.

Apis Mellifera. The Swedish botanist and zoologist Carl Linnaeus introduced the system of nomenclature, for identification of life. Linnaeus gave the honeybee the name *Apis mellifera* in 1758 and it appears in his book "Systema Naturae" (10th edition). "*Apis*" is the Latin word for "bee", "*mellifera*" comes from the Greek "melli", honey, and "ferre", to bear - hence, the scientific name means the honey-bearing bee. Three years later in 1761 Linnaeus realised his earlier mistake that honeybees bear nectar not honey, so he attempted to change the name to *Apis mellifica* - the honey-making bee. However, according to the International Code of Zoological Nomenclature, the older name has precedence (the Principle of Priority, first formulated in 1842), so we have continued to use the name *Apis mellifera*.

How not to get lost

We all seem to know how to programme the Satnav systems nowadays, but scientists are becoming increasingly alarmed today, as we seem to have lost the capability to read and use a map and compass. How has life on earth managed to work out how to navigate in its own unique environment, after all they do not read a map or use a compass, or do they?

Each environment produces its own problem for the inhabitants such as darkness at night or living under the sea, cloudy days, winds and currents, when they are navigating to find food, mates or to return to their nesting ground.

If we look at the main ways navigation is used by life on earth some species navigate across the world and back again, often in hostile conditions.

A list of some of the navigational methods is set out below.

Sun navigation: this only works during a cloudless day. As the sun moves across the sky from east to west during the day, an angular difference occurs called an azimuth. It simply measures the progress of a heavenly body from east to west by reference to the (imaginary) moving point on the horizon vertically beneath it. So an azimuth of 090 degrees is due east, 180 due south.

This factor must also be taken into consideration and calculated by the user. Its position also varies in the sky throughout the seasons, being directly overhead in midsummer and inclined to the horizon in mid winter.

Polarised light: this can be seen by insects and some birds, and will penetrate through clouded skies.

Sound navigation under the sea: used by mostly marine dwellers as water conducts sound waves four times better than air.

Smell navigation: used to locate food and mating areas.

Star navigation: the Milky Way on a clear night acts rather like the sun; the polar star is another location point.

Dead reckoning: estimating method by memorising of the route.

Landmarks: using geographical features, such as hills, mountains and lakes, hedge and tree rows

Sonar navigation: best known in bats and dolphins.

Electrical field navigation: found in some fish and sharks.

Magnetic navigation: some animals and insects have sensors in the body to pick up the earth's magnetic fields.

Map and compass navigation: the human preferred method, but some animals also have a memory map and an inbuilt magnetic compass.

So what does the bee use to navigate by?

The workers, drones and queen all make journeys outside the hive too far off places so they must all be able to navigate. The famous waggle dance tells the workers how to find the food crop that has just been brought in; this tells the worker the distance and whereabouts of the food by using reference to the sun and polarisation orientation in the sky, allowing the next bee to memorise the data. An elementary form of map making and reading.

The honeybee can detect the earth's magnetic field by using microscopic crystalline chains of magnetite crystals in its abdomen, but it may not make direct use of it for navigational purposes. More probably, it uses the regular daily changes in the earth's magnetic intensity that occur around sunrise and sunset to calibrate the internal clock that governs its sun compass.

The bees have ocelli on top of their heads. These are not used to see with but can detect the polarisation in the sky; they also locate the sun's position visual.

Bees rely on scents to find their food and use smell navigation when approaching large groups of flowers. Drones also use it when locating a queen during the mating flights. However, the odour plumes they follow are diluted very rapidly as they disperse. They may initially be responding to a single scent molecule, but the plumes often break up completely in a moving airstream.

The brains of bees contain two structures that seem to be of great navigational importance. The so-called mushroom body stores long-term memories based on smell and sight, while the "central complex" controls the course that the animal follows, in many cases making use of skylight polarization patterns to do so. Because these structures are shared so widely with other insects, it is thought that they must have emerged at a very early stage in the evolutionary process. Exactly how the animal chooses which way to go and initiates the appropriate movements is still a mystery, but interactions between the mushroom body and the central complex seem to play a crucial part in the process.

The book stated below by David Barrie inspired this article; it helps explain the wonderful world of navigation and answers some of the mysteries surrounding them. Highly recommended.

INCREDIBLE JOURNEYS: Exploring the Wonders of How Animals Find Their Way, by David Barrie. Published by Hodder & Stoughton. Parts quoted with kind permission.

Which side of the fence do we sit?

The more senior member will remember his or her first summer days of driving the car, when you had to stop to clear the windscreen from dead flying insects almost every hour! What has happened? Where have they all gone? This anecdotal evidence shows something is amiss!

The UK is proposing a new farming policy for when we leave the EU. It is being passed through Parliament now; the farmers think it needs more adding to it and they are criticising the insect policy included. This is very good for our insects, wildlife, as it will mean more (in a nutshell) hedges, wildflower meadows, and fallow fields. Farming policy since the Second World War has put a lot of emphasis on simply producing more food in the UK, it currently supplies about 60 per cent of our needs. Innovative farmers are already cutting their use of chemicals. They are finding they can cut costs and produce healthy crops

Fig 23. A fence sitting worker bee!

by working with nature. However, the sobering news is that neonicotinoids are not the only pesticides affecting pollinators. The reality on our farmland is that wildlife is routinely exposed to a whole cocktail of chemicals. Evidence shows that fungicides, used to control disease rather than pests, could increase the toxicity of neonicotinoids to bees.

Have you stopped to consider broad-spectrum herbicides like glyphosate, which are not selective about the weeds they kill – so they reduce the availability of pollen and nectar from wild plants? The RSPB has commented, "Our beleaguered farmland birds have declined by 56 per cent between 1970 and 2015 along with declines in other wildlife linked to changes in agricultural practices, including the use of pesticides." Therefore, we know that pesticides are harming our birds and bees. 10 years ago, the hedgehog population was 30 million; now there are less than 10 million. Farmers must protect their crops from pests and diseases but current levels of pesticide use are unnecessary as well as damaging.

A report from Germany on a study that took place over 27 years is abridged below; it sets out the grim reality.

"Global declines in insects have sparked wide interest among scientists, politicians, and the general public. Loss of insect diversity and abundance is expected to provoke cascading effects on food webs and to jeopardize ecosystem services. We measured total insect biomass using Malaise traps, deployed over 27 years in 63 nature protection areas in Germany (96 unique location-year combinations) to infer on the status and trend of local entomofauna. Our analysis estimates a seasonal decline of 76%, and mid-summer decline of 82% in flying insect biomass over the 27 years of study. We show that this decline is apparent regardless of habitat type, while changes in weather, land use, and habitat characteristics cannot explain this overall decline. This yet unrecognized loss of insect biomass must be taken into account in evaluating declines in abundance of species depending on insects as a food source, and ecosystem functioning in the European landscape."

The full open access article by CA Hallmann can be found here: *www.journals.plos.org/ plosone/article?id=10.1371/journal.pone.0185809*

All insects are of major importance to the ecology, however we need to sit down firmly on one side of the fence and decide to change/challenge things; some of this is being addressed now, but beekeepers need to start thinking smarter and to lobby our MPs on the particular issue of honeybees. Consider the following: the registration of all hives; setting up a UK-based queen rearing programme and banning imported queens; stopping hive products coming in without being analyzed for disease and pests; keeping local bees that are best suited to their environment; allowing native European black bees to be kept in controlled areas.

A final thought on breeding. One common consensus is that we all want gentle bees and want to breed for this but I have a hive that is very defensive this year – the bee's words not mine - I would describe them as very aggressive! Now will this trait be an advantage to them when the Asian hornet arrives? They were a strong colony with no wasps around them this year either, unlike my other two hives. We are tinkering at all levels. Honey and wax are a luxury. It has been estimated that honeybees pollinate about 30% of human food crops and are the fourth most important animal to humans!

Look to your drug information leaflet about all the side effects; there is always a trade-off here in order to get better. We treat bees with chemicals but do not know all the side effects; science is only now understanding the major importance of the human gut biome - what does medication do for the bees' gut for instance? They suffer from bacteria, viruses and moulds like us and are marvels in their own world, using senses we do not have; how do all these treatments affect their senses?

Ponder at will but do not take too long. I am not against some methods used but we must start making informed choices in the interest of the non-human world. Over to you neighbour.

Number 30, the bee's special number?

Have you ever wondered why you have to take 30 bees for sampling when checking for disease or wing morphology?

It is all about '*Sampling and predicting*', some crystal ball gazing with some acquired knowledge! Counting all the bees in a hive or the number of fish in a shoal is not a very practical means.

The method we use is called proportional reasoning.

If you want to know the number of bees in a hive, take a small sample of about 30 bees. Mark each bee in the same way you would mark a queen, and then return these marked bees to the hive. After some time, when you think the marked bees have mixed with the other bees in the colony, then take another sample.

Let us assume this sample contains 100 bees. In this sample, there will likely be some of the marked bees from the first sample. Suppose there are six marked bees. In effect, you have the following proportion:

Number of marked bees ÷ Number of bees in sample = Number of marked bees ÷ Number of bees in hive

Using the above example, we get $6 ÷ 100 = 30 ÷ x$, where x is the total number of bees in the hive.

Solving the above equation, we get $x = 500$. Therefore, from our sample we can predict that there are 500 bees in the hive. We have to sample a lot more bees in a full hive in the summer, due to the sheer numbers involved, when there are up to 50,000 in residence.

Timing is of importance

A honeybee colony is a complex Superorganism with changing features in response to seasonal changes in the environment. The average age increases, for example, in colonies in the autumn in temperate regions, because of the transition to winter bees. Immediately after a colony has produced a swarm, the bees remaining in the nest will have a large proportion of bees younger than 21 days, lowering the average age of bees in the colonies.

The same is true for recently caught swarms, because the brood will not have had enough time to develop, and one could expect rather an over-aged structure. Therefore, it is recommended that your aim is to have an average / normal / representative sample, by allowing time for the colony to adjust before collecting them.

In most cases, you want older flying bees that have been exposed to all the problems in the hive.

Disease Methology

In the excellent 'Beekeeping Study Notes - Microscopy certificate.' by the Yates team , in Appendix 1 they have set out the mathematical formula for disease sampling, which fortunately for us without the university training, they have simplified and have also included a graph for easy calculations.

The basis of this chart allows for the number of infected bees in the 30-sample rate to be predicted. For example if three bees where found infected out of a sample rate of 30 then we can assume that there is a possibility that 99% of that colony has a 47.5% of its population also infected. If only one bee is found infected then there would be a possibility that 99% of that colony has a 20% of its population infected.

Yates could not find any recent reasons for sampling when writing the study notes in 1995, so set their own equations up; these seem to have stood the test of time.

There has now been an update re how, when and where statistical analysis can be used for bee disease research; please look on line for 'The Journal of Apicultural Research, Coloss Beebook volume 2'. I warn you it is heavy going!

I must also point the avid reader to the late Dave Cushman's beekeeping web site. *www.dave-cushman.net* where he, as usual, explains things in great clarity.

An alternative use of old wings

Bees' wings, a flimsy floating crust of tartar formed in Port and other wines after long keeping.

So now, you know where old wings go, no chance of this happening in our household!

Tartaric acid is the main organic acid that occurs naturally in many fruits, most notably in grapes. Its salt, potassium bitartrate, commonly known as cream of tartar, develops naturally in the process of winemaking. It is commonly mixed with *sodium bicarbonate* and is sold as baking powder that is used as a leavening agent in cake making. The acid is also added to foods as an antioxidant and to impart its distinctive sour taste.

Who will be the bees advocate?

What do honeybees want and need and how do they make their decisions? What do we really know about the wild bee's life style? The arrival of the framed hive has allowed honey to be the main centre of beekeeping activities, together with an opportunity to study bees close up, albeit in an artificial home. The BBKA has an excellent structure to learn about beekeeping. However new evidence is coming to light and not being validated and rolled out as quickly as perhaps it could be.

Like any subject, there is differing opinions of the best way to keep honeybees. I think we all share a common view but we need to rethink some of the methodology we practice in beekeeping. No one group has a monopoly of knowledge. The threats over the last 30 years have radically changed the methods of keeping and treating bees.

As an example, commercial beekeeping in the USA which uses thousands of colonies for pollination of orchards is not the way forward; it has been proved a monoculture of pollen does not provide the variety of proteins and minerals for their needs, hence one of the reasons for colony collapse, at a rate of 40% losses every year. China now has to pollinate by hand in several regions. We need to feed the world and not all ideal methods can be used, so we need to look at everything and do our bit locally where we can make a difference and try to help in other areas where modern farming will have to take precedent.

New science shows that climate change, pesticides, herbicides, removal of hedgerows and farming methods are affecting all species worldwide. Residue chemicals in wax are altering bee's fertility and development.

We must allow natural selection to take place; the *varroa* mite has already become resistant to treatments.

Prof Seeley has shown that it will take about 5 years of major losses before resistance does have an effect. We need to look at the longer view.

Some of the things we do in our interest, such as importing queens, which are not suitable for our local environment, have resulted in failed colonies.

Killing the queen when the colony exhibits undesirable traits, such as defensive behaviour, this could possibly be an aid in balling and killing the Asian hornet.

Selective breeding for whose benefit? If you select for a trait you often affect other traits to the detriment of the colony. Honeybees have a unique mating system, not seen in any other creatures on the planet.

In his book 'The Lives of Bees' Prof Seeley has studied wild bees over a period of 30 years to see how they live, with a view of managing bees to reflect their wild condition; he has made a comparison chart of the differences. This summary has been taken from his book with his kind permission.

	Wild colonies	Framed colonies
1	Colonies are genetically adapted to their location	Colonies are not genetically adapted to their location
2	Colonies live widely spaced in the landscape	Colonies live crowded in apiaries
3	Colonies occupy small nest cavities	Colonies occupy large hives
4	Nest cavity walls have a propolis coating	Hive walls have no propolis coating
5	Nest cavity walls are thick	Hive walls are thin
6	Nest entrance is high and small	Nest entrance is low and large
7	Nest has 10%– 25% drone comb	Nest has little (< 5%) drone comb
8	Nest organization is stable	Nest organization is often altered
9	Nest-site relocations are rare	Hive relocations can be frequent
10	Colonies are rarely disturbed	Colonies are frequently disturbed
11	Colonies deal with familiar diseases	Colonies deal with novel diseases
12	Colonies have diverse pollen sources	Colonies have *homogeneous* pollen sources
13	Colonies have natural diets	Colonies sometimes have artificial diets
14	Colonies are not exposed to novel toxins	Colonies exposed to insecticides and fungicides
15	Colonies are not treated for diseases	Colonies are treated for diseases
16	Honey not taken, pollen not harvested	Honey taken, pollen sometimes harvested
17	Combs not moved between colonies	Combs often moved between colonies
18	Honey cappings are recycled by bees	Honey cappings are harvested by beekeepers
19	Bees choose *larvae* for queen rearing	Beekeepers choose *larvae* for queen rearing
20	Drones compete fiercely for mating	Queen Breeder may select drones for mating
21	Drone brood not removed for mite control	Drone brood sometimes removed and frozen

It would seem that we are a long way away from keeping hived bees in tune with their preferred style of residence.

The Darwin way of keeping bees is the last chapter of his book, giving many recommendations of how we can still keep bees but to their advantage. This proposed method is

Fig 24. Will you be my advocate?

mainly aimed at the hobby beekeepers that make up a substantial percentage of the beekeepers in the world.

A more bee orientated method will change and challenge the way we hobby bee-keepers will keep bees and will turn some methods used to manage colonies on their heads, such as ventilation, as insulation and surface texture all affect condensation and humidity.

The beekeeper does not seem to have helped the bees at any stage of their lives; we manage them to suit ourselves, and are responsible for most of their current and past woes.

A must read for all beekeepers who are looking to help their bees.

'The Lives of Bees', Prof T Seeley, ISBN-10: 0691166765

A final thought from Prof Leslie Bailey's book, Honeybee Pathology.

Beekeeping today is still as it has always been the exploitation of colonies of a wild insect; the best beekeeping is the ability to manage them naturally and at the same time to interfere as little as possible with their natural propensities.

Who will listen to them and be their advocate?

Record Breaking

In 1983, a scientist named Spangler published a paper about one of the beekeeper's nemesis, the Greater Wax moth, *Galleria mellonella*. The main claim to fame was its ability to hear ultrasonic frequencies up to 320,000 vibrations per second (320 kHz); the highest frequency sensitivity of any animal.

The moth would respond at this frequency by folding its wings back and falling to the ground in defence.

Fast forward to 2013 and the University of Strathclyde published their own data.

A possible evolutionary war is being played out between the bats, which use ultrasonic calls to locate insect prey. (The highest known frequency of bat echolocation is 212 kHz) and the greater wax moth that uses a tympanic hearing organ on their abdomen consisting of four receptor cells, to listen for the approaching bat.

Why they have a much higher hearing level than their predators is possibly due to evolutionary pressure aeons ago, perhaps for predators with super echolocation.

The male moth also uses high frequency pulses of sound in their courtship calls.

In comparison, young humans are able to hear up to 20 kHz and honeybees can generate sound up to 1000 Hz but can only hear up to 500 Hz.

Just remember in the wild the Wax moth perform a service for the feral bees, by clearing out old colony sites in trees, in preparation for a new swarm in the future.

For the scientists amongst you, the researchers used two different experimental methods: laser Doppler vibrometry to record the tympanal membrane mechanics, and electrophysiology to record the neural response of the auditory nerve. Both experiments were done 'separately or simultaneously' to record the mechanical response of the membrane and the neural response of the ear. Neither method involved any contact with the moth's abdomen.

Revenge?

I noticed over sixty mustard coloured oval spots about 1 mm in size on my car when passing by, early in the morning. They looked suspiciously like those seen on the outside of the hive, when bees have a problem, such as dysentery.

If a bee is unable to void the contents of its distended rectum for an extended period of time, such as in the winter, or due to bad weather, indigestible foods like starches may ferment due to *bacteria*, *yeasts* and *micro fungi* causing *dysentery*.

The day before I had a quick look inside the hives for the first time this year and collected 30 bees, mainly from outside the hive, as I leave a seed tray underneath the entrance in order to keep an eye on what is happening without looking inside. I went on to check for *Nosema* and *Acrine* mite. The good news, none was found in either of my hives.

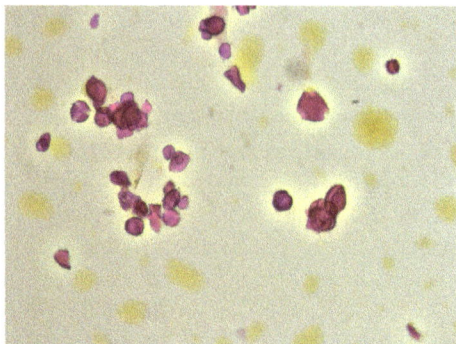

Fig 25. Mustard spot evidence.
Pollen stained red also showing
the yellow wax from their outer
coating. X 200 magnification.

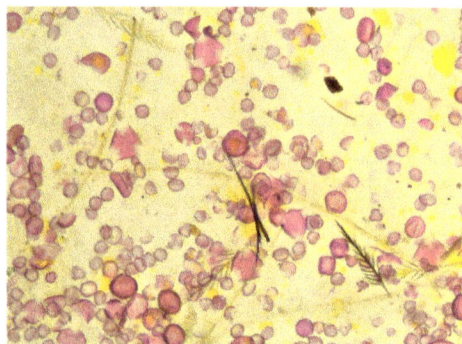

Fig 26. Comparative evidence. The
Nosema sample (none evident). Taken
from the ground up abdomens, showing
pollen and hairs. X 200 magnification.

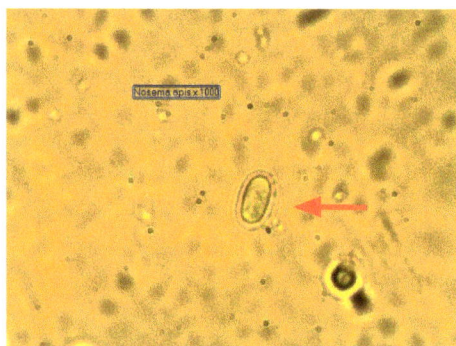

Fig 27. A sample with Nosema, small rice
shaped particles, black arrow and mixed
pollen grains. X 1000 magnification.
Open Government Licence. Beebase.

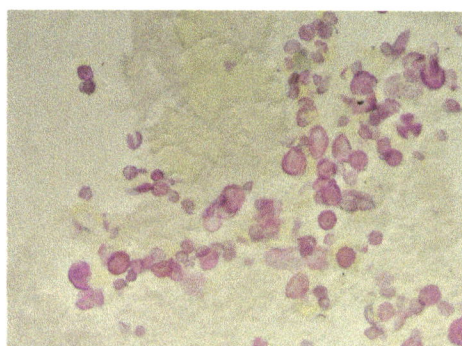

Fig 28. Larvae cut contents.
X 200 magnification.

Were they seeking revenge I wondered? Butterflies were not in evidence en mass, so they were excused from being the culprit.

I decided to scrape some of the spot materiel off the car and put it into some *alcohol*, and then made a slide of it with some added stain to help show up the contents. The evidence was conclusive. Guilty as charged. Bee poo!

I inadvertently pulled apart some grubs when splitting the super, so I took them away to look at them under the microscope. From the squishy ones I took a sample from their gut contents, a mustard coloured liquid, to see what they had been fed. The *larvae* where about 5-7 days old, so fully formed.

The slide revealed pollen, which *larvae* need a lot of as it is their only source of *protein*, *fat*, *vitamins* and *minerals*, which is digested and reused to fuel their massive growth over the first 7 days of development.

Faecal matter includes pollen *exines*, the outer hard coating, the inside cytoplasm being digested, pollen *lipid* globules, the yellow matter in the pictures. Sloughed *epithelial* cells from the *Ventriculus*, which covers the pollen, mass during digestion; this is thought to protect the gut lining from transit damage due to the rough outer *exine* surfaces.

So there you have it, a potted history of bee poo!

Slovenian Beekeeping

On a recent holiday during May several years ago, we visited our good friends Pam and Alan who live in Slovenia; they very kindly provided us with the opportunity to visit the home of their friend who is the local English teacher, Zorka and whose husband, Branco just happened to keep twenty hives in their mountain meadow! Being a novice beekeeper, this gave me a wonderful chance to see beekeeping from a continental viewpoint. Our host had done her homework; she had looked up a few beekeeping terms in English as her husband only spoke four local languages!

It would seem gestures and a little knowledge goes a long way when trying to communicate bee matters.

They live in the Julian Alps mountain region (1400m) which has deep snow every winter and in order to protect their hives from the cold and wild bears they have chosen to keep them in a traditional shed like structure, which also doubles up as a drying-storage rack for hay, beside and above the building.

The front area has a shutter on for shade in the summer and insulation in the winter; inside the hives are built side-by-side and stacked on top of each other; the fronts are painted different colours to aid the bees in recognising their own home.

Fig 29. Side profile of a typical hive set up.

In the olden days, the beekeeper would paint a picture on each entrance, which would give the bee a 'pattern' to see as they approach the hive. These doors are now for sale in the antique market in the capital.

Once inside the bee house each hive can be accessed through an open door, which has a removable mesh panel to stop the bees from entering the room.

Fig 30. The front with the main door acting as a shade.

The method of bee keeping is the same as in England except things are done through the side entrance; one advantage is that you can see inside without disturbing the bees.

The plastic container on the upper right of the picture is for *Varroa* treatment; the Slovenes tend to put in larger celled foundation to encourage drone brood all over the frame which they then remove when full; not wishing to waste anything, it is given to the hens, whom seem to enjoy the extra protein dish!

Like us, they practice a variety of methods to control diseases.

From twenty hives, he extracted over 500 kg's (averaging 25 kg or 55 lbs per hive) of honey last year, this being one of the better seasons. Branco's yield does vary and winter loses are low, however due to legislation they cannot import any bees and must breed only Slovenian bees which are of the Carniolan strain (*Apis mellifera carnica*). Their one weakness is that they have a propensity to swarm regularly; he said that he only just lost a swarm the day before. We witnessed a swarm a day later where we were staying, making a new nest in the neighbour's chimney. One of their advantages is that they can over winter well on limited stores, which suit the local climate conditions; he does feed sugar after taking off the surplus honey.

In general, there has been a decline in bees locally, but with the larger land mass and abundant wild flowers together with the small scale farming methods; they have been able to maintain stocks so far.

There are two million Slovenians living in an area about the size of Wales, which gives a density of people per square mile of about four times less than in England.

Fig 31. Antique brood chamber door.

Fig 32. The front entrances, fully opened.

Fig 33. The inside hive exposed.

Fig 34. Typical Slovenian meadow.
All pictures by Catherine Kingham.

Granola and honey

For my sixtieth birthday present our son and his partner took my wife and I to New York. We ate out for breakfast several times to experience the American way, forget the eggs over easy and fried potatoes with onions, go for the healthier option of yogurt topped with a ring of granola with added fresh fruit, such as blue berries and strawberries or pears in the centre, plus the extra ingredient, some runny honey drizzled either around the yogurt or over the fruit.

Inspired by the meal I returned home and looked up several recipes for granola, so after several very pleasant trials, I came up with a proven recipe that we all liked.

300gm plain muesli base or porridge oats

2 tablespoons sunflower oil

2 tablespoons runny honey

125 ml maple syrup

50 gm sunflower seeds

50 gm linseed or sesame seeds

1 teaspoon vanilla essence

1 teaspoon finely crushed cardamom seeds

1 teaspoon ginger powder

1 teaspoon nutmeg

½ teaspoon cinnamon

100 gm dried coconut

Generous pinch of salt

100gm Dried fruit (sultanas, raisins and cranberries or your choice)

Mix together the oil, honey and maple syrup, stir then add the seeds, salt and spices, mix well then add the muesli, stir to make sure that the cereal grains are covered then transfer into a shallow baking tray spreading evenly.

Bake in an oven at 150ºC electric, 130ºC fan oven or gas mark 3 for 15 minutes, remove from the oven and stir in the fruit and coconut returning to the oven for about 5 minutes then stir, return to the oven for about another 5-10 minutes, stirring if needed, checking it has become dry and slightly coloured, this depends on the depth of mix.

Allow to cool, then store in an airtight container, it will keep for up to 3 months.

An opportunity to see in ultra violet!

As a novice bee keeper who recently attended a beginner's course run by Kay and Julie from the North Devon Beekeepers Association, we were required to read a book every week from the library; one such tome that interested me was called 'The Buzz about the Bee. The Biology of a Superorganism' by Jurgen Tautz. Published by Springer.

This book tells the story of the honeybee, based on recent scientific observation; one of the chapters was devoted to sight.

Where I work, they use ultra violet light at 365 nanometres to check on component part (the other higher harmful ranges of ultra violet light are filtered out). This gave me an idea; why not look at some of the flowers that bloom throughout the season to get a bee's eye impression! The thought of photographing them as well came to mind, but on reading up on this matter it requires some serious equipment and photographic knowledge, however there are some excellent sites displaying the 'patterns' using ultra violet light photography on the internet.

In order to try to understand this other 'spectrum' and how it affects the bee's perspective of vision I have done some research to help clarify UV electro magnification.

Scientists have divided the ultraviolet part of the spectrum into three regions - the near ultraviolet, the far ultraviolet, and the extreme ultraviolet; the three regions are distinguished by how energetic the ultraviolet radiation is, and by the "wavelength" of the ultraviolet light, which is related to energy. The name comes from the Latin *ultra* meaning 'beyond' and violet being the colour of the shortest wavelengths of visible light at 400nm. UV light has a shorter wavelength than that of violet light.

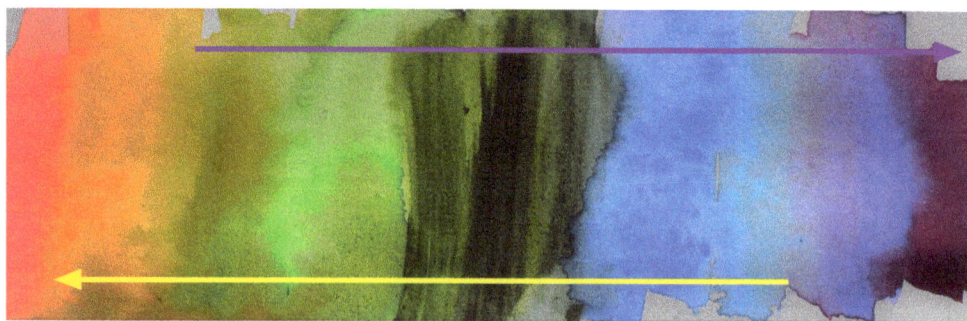

Fig 35. Starting with infra red and finishing with ultra violet light, the purple arrow shows what colours a bee see and the yellow the human's range of vision.

The sun emits ultraviolet radiation in the three bands; however, the Earth's ozone layer blocks 98.7% of this UV radiation from penetrating through the atmosphere.

The UV range of the spectrum has no predefined colours, so we are free to assign any colour we like, albeit that under an ultra violet light, depending on the frequency of the lamp used and any filters in place, the colour ranges from pale violet to almost black;

so do not compare colours, compare patterns; fluorescence may be a common trait to most flowers, but might be of temporary occurrence for parts of the flower.

Anthers, style, and pollen grains often scattered on the petals occasionally are seen to fluoresce. Strong fluorescence has been noted from the nectar glands. Fluorescence from outside the bracts is exhibited by some species.

However, not all flower species have the typical centric UV pattern, which may be confined to symmetrical petal flowers, these might act to attract the bee to the centre of the flower and make it stand out from its surroundings; some flowers exhibit a virtually endless variety of spectral signatures.

Scientists in the Ecology of Vision group at the School of Biological Sciences (University of Bristol) have argued that much of our understanding of how insects and birds see UV colour is fundamentally flawed.

If you watched a wildlife series with, say, the red light source of your television removed (or if you were red-green "colour-blind) and you then came up with conclusions about colour variation in the natural world, would anyone believe you? Probably not, but then that is what we humans are doing every time we think we are seeing the colour world of non-human animals. Unlike other variables such as length, width, mass, or time of day, colour is not an inherent property of the object; it is a property of the nervous system of the animal perceiving the light.

Understanding Colour Vision in Humans

The sensation of colour stems from the differential stimulation of the different types of photoreceptors in the retina. Each cone type produces an output, and it is their differences in output at a particular point on the retina that underlies the sensation of colour. In humans, there are only three types of cones, absorbing photons in different regions of the spectrum. Due to the appearance to humans of monochromatic light at these wavelengths, these three cone types are called "red", "green" and "blue" respectively. Consequently, for humans, all hues can be produced by mixing red, green and blue light. This is how a colour television set works; a mixture of three wavelengths produces several million apparent "colours". However, there are several problems. Firstly, different wavelength spectra can produce the same hue; as long as the output from the three types of cone remains the same, the hue is the same. Secondly, the same wavelength spectra will produce different hues to animals that differ in the absorption spectra of their cone types. Thirdly, humans have trichromatic, three-dimensional, colour vision because we have three interacting cone types. Animals with two interacting cone types, such as most mammals other than old-world primates, have two-dimensional colour vision (similar perhaps to the faulty colour TV set mentioned earlier). It is harder to imagine what colour vision with more dimensions than three might be like, but animals with four and five dimensional colour visions exists. When asked to identify the colour of an object, you will most likely speak first of its hue. Quite simply, hue is how we perceive an object's colour, red, orange, green or blue. Chroma describes the vividness or dullness of a colour. In other words, how close the colour is to either gray or the pure hue. For example, think of the appearance of a tomato and a radish. The red of the tomato is vivid, while the radish appears duller. One interesting factor for me is although I am not colour blind I do have astigmatism in one eye and I perceive colour differently in each eye, one being a shade darker than the other is, but as they work as a pair, my good eye over-rules the other.

The Colour Vision in Bees

Bees, like humans, have three receptor types, although unlike humans they are sensitive to ultraviolet light, with loss of sensitivity at the red end of the spectrum. This spectral range is achieved by having a cone type that is sensitive to UV wavelengths, and two that are sensitive to 'human visible' wavelengths. Remember, because 'colour' is the result of differences in output of receptor types, this means that bees do not simply see additional 'UV colours', they will perceive even human-visible spectra in different hues to those which humans experience. Fortunately, as any nature film crew knows, we can gain an insight to the bee colour world by converting the blue, red and green channels of a video camera into UV, blue and green channels. Bees are trichromatic, like humans, so the three dimensions of bee colour can be mapped onto the three dimensions of human colour.

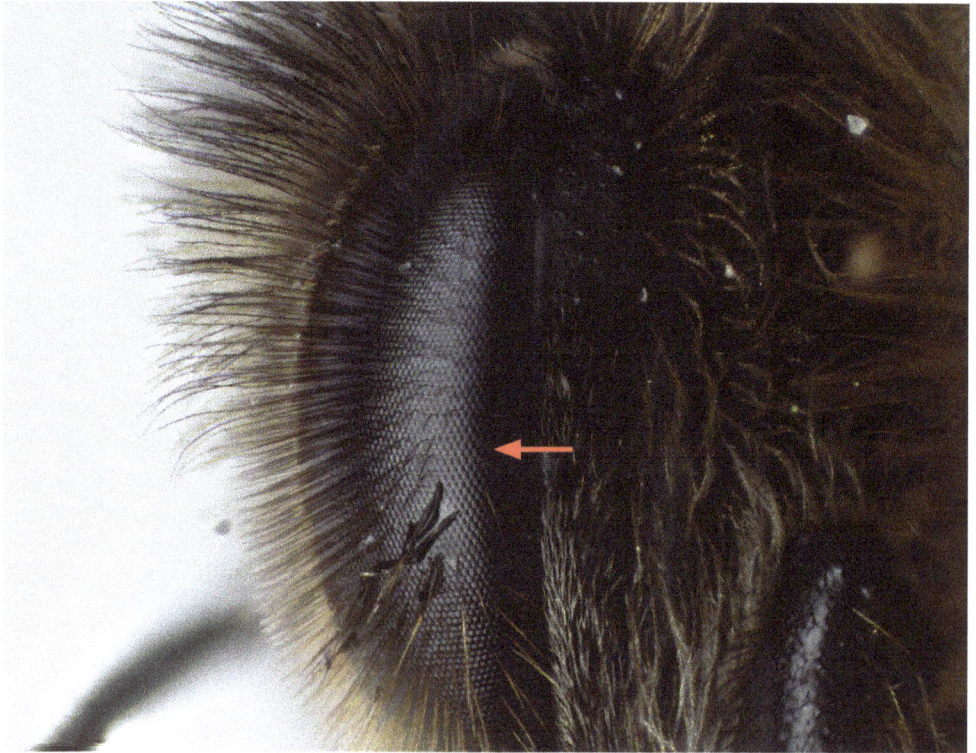

Fig 36. The compound eye of a worker bee, showing the hexagonal facets.

Until relatively recently it was thought that humans had amongst the best colour vision of any animal, and that most animals' spectral sensitivities lay within the human-visible spectrum. This misapprehension persists outside the visual sciences. Other creatures can see in four or more primary colours.

One of the advantages of why bees and birds can see in UV light is because many fruit, flowers, and seeds contrast with their background much more strongly in UV than human-visible wavelengths. As insects and plants have devolved together in evolutionary time, they have built up many symbiotic methods between them.

Human and bee comparative colour chart

What we see	What bees see	Add in UV
Red	Black	UV Purple
Orange	Yellow-Green	
Yellow	Yellow-Green	UV Purple
Green	Green	
Blue	Blue	UV Violet
Violet	Blue	UV Blue
Purple	Blue	
White	Blue-Green	
Black	Black	

It has been suggested that flowers have developed colour as a means of standing out amongst the surrounding green foliage as an aid to attract insects for pollination, with the added bonus of an ultra violet central pattern to indicate the pollen area. Bees are also guided by smell, but that is a subject for another article, which is beyond my knowledge.

Fig 37. Dandelion (Taraxacum officinale) as humans perceive it and as a bee.
By kind permission of István Bocskai: www.momentslumiere.com.

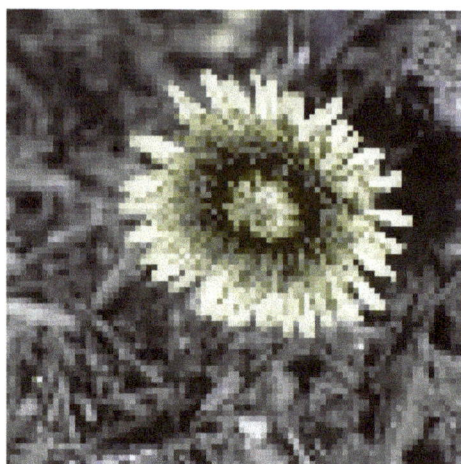

Fig 38. How a bee might 'see' when approaching a flower with its compound vision.

At sleep or just resting?

Do bees sleep? The answer appears that they do.

Sleep is a dynamic phenomenon that changes throughout an organism's lifetime, relating to possible age or task associated changes in health, learning ability, vigilance and fitness. Sleep has been identified experimentally in many animals, including honeybees (*Apis mellifera*). As worker bees age they perform different tasks, such as cell cleaners, nurse bees, food storers and foragers. These four tasks were used as a basis for a recent study. Performing differentially could influence the duration, constitution and periodicity of a bee's sleep. The study looked at the duration and periodicity of sleep when bees were outside comb cells, as well as duration of potential sleep when bees were immobile inside cells. All four-worker bee common roles examined exhibited a sleep state.

As bees aged and changed tasks, however, they spent more time and longer uninterrupted periods in a sleep state outside cells, but less time and shorter uninterrupted periods immobile inside cells. Cell cleaners and nurse bees exhibited no sleep - wake rhythmicity, food storers and foragers experienced a 24-hour sleep - wake cycle, with more sleep and longer unbroken bouts of sleep during the night than during the day.

If immobility within cells is an indicator of sleep, the study reveals that the youngest adult bees sleep the most, with all older bees sleeping the same amount. Several 'sleep signs' are deemed critical by most researchers when defining sleep behaviourally in animals; these include a sleeping organism exhibiting a specific posture during easily reversible bouts of relative immobility, during which its arousal threshold is increased; such a state should be internally controlled. Worker bees' sleep signs include antennal states associated with sleep in bees as either, antennae motionless, slightly twitching, or exhibiting larger, usually swaying motions.

Fig 39. Worker honeybee at rest on a wall.

Patterns of sleep and immobility were consistent across the studies, with bees sleeping more outside cells when older, and spending more time immobile inside cells when younger. This increase in sleep outside cells with respect to age/caste held true for total antennal immobility, a state correlated with high arousal threshold and speculated to be the deepest state of sleep. Cell cleaners and nurse bees exhibited more sleep and more deep sleep outside cells when they became food storers and, in the case of variable antennae, again as food storers became foragers. As younger bees aged and changed tasks, they also experienced longer unbroken bouts of sleep outside cells, increasing as nurse bees became food storers, and again as food, storers became foragers, but experienced shorter bouts of immobility inside cells. Since 1952, several studies have been concluded about honeybees sleeping patterns; this study was done in 2008.

This article has been summarised from the study by the following authors. Barrett A. Klein1, Kathryn M. Olzsowy, Arno Klein, Katharine M. Saunders and Thomas D. Seeley 'Caste-dependent sleep of worker honeybees'. The study has also put four short videos on the web showing the stages of sleeping workers.

Plants

Leading the way

On visiting Sidmouth in early August, I was amazed to see that the local town council had allowed many wild flower patches to be created all across the town centre, to encourage pollination by wild insect. These require little maintenance, provide an eye catching display and they have used different seed mixes throughout each bed.

Next step? Please lobby your local council!

Fig 40. A typical wild flower bed. Picture by Catherine Kingham.

The Hollyhock's spiky secret

The majestic Hollyhock feature in many English gardens; their stately blooms can reach over 2 metres in height and come in many different colours. The plant originates from southwest China in the 15th century. It relied on the local animals and insects to pollinate it. However, in England they are absent so what are the problems? Very large spiky pollen apparently. Their pollen grains are one of the largest, at 130 microns with about 150 spikes on the outer surface, counted using a scanning electron microscope; the average size pollen is between 25 and 40 microns. The honeybee has specialised plumose hairs, a single stem with branching hairs about 40 microns apart, which help trap pollen between them. *Althaea rosea*, the common hollyhock, is an ornamental plant of the Mallow family; its Latin name is *Malvaceae*. It flowers from June to September and offers a food source for bumblebees and honeybees, although they do not make honey from its nectar.

The outer and inner surfaces of pollen grains (*exine* and *intine*) have different structures together with differing shapes and sizes between plant species. These features enable them to be identified with a compound microscope at 400 times magnification. Modern DNA methods are now available and offer a definitive answer as it takes time to learn how to master their identity. Why does pollen grain have different structures? Possibly to stop cross-pollination and to target its own species. Bear in mind that bees and flowers have co-evolved, nectar attracts them as a sugary reward, as part of the process; pollen is attracted to their bodies by static electricity, pollen kit, a sticky outer surface coating and physical contact. Nectar provides carbohydrates and pollen is protein. Stop there! The deal is

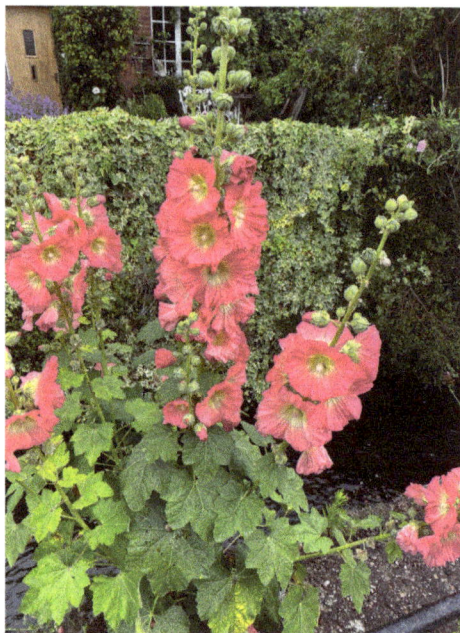

Fig 41. Majestic Hollyhock flowers in our village. Picture by Catherine Kingham.

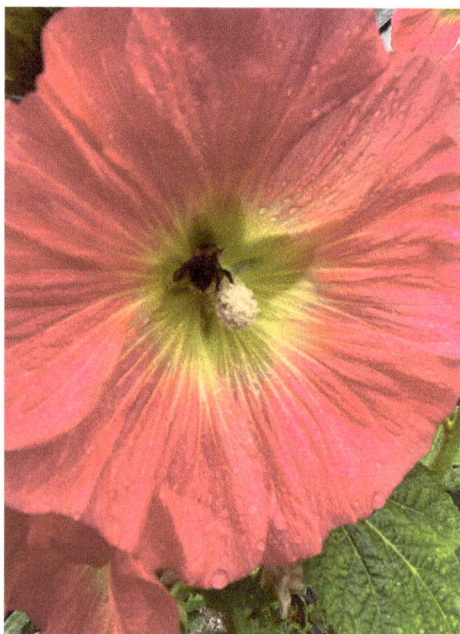

Fig 42. Bumble bee visiting the nectaries. Picture by Catherine Kingham.

to transplant pollen to other plants of the species to help cross-pollinate them, surely, not to eat it. Some bees have a special set of hairs on their blacklegs, the *Corbicula* or pollen basket; they comb the pollen backwards with their other legs then mix it with saliva and press it between the special presses like toothpaste onto the outer stiff hairs.

Has the Hollyhock developed an armoury to prevent this theft, which cost the plant a lot of energy to produce and reduces it reproductive success?

The Buff tailed bumblebee seems to be one of its most frequent visitors in the UK. Due to the larger sized pollen it adheres readily to any bee's hairs all over their bodies, but due to the size and spikey outer surfaces of the pollen, the bees are not able to compress it, so must sift it out when packing pollen onto their legs. It would seem the plant has a mechanical defense against being eaten, allowing more pollen to do its job of fertilising other Hollyhocks.

The spines reduce the contact zone between butting pollen grains, because only one or a few spines interact with the pollen wall of a butting pollen grain. The large size of the pollen grains may also contribute to the effect of a reduced contact zone. Plants have also developed other means of protecting pollen, such as bitter tasting and toxic coverings of their outer coating. As honeybees taste through their feet, they can recognise them, preventing them from feeding poison to their *larvae*. It might also indicate that the surface spikes will harm the *larvae*'s intestines, so they selectively choose not to harvest it. The honeybee when eating pollen, wraps a cell membrane round the pollen as they enter the true stomach, the *ventriculus*. It is thought that this is done to protect the internal lining from the outer rough, surfaces of the pollen.

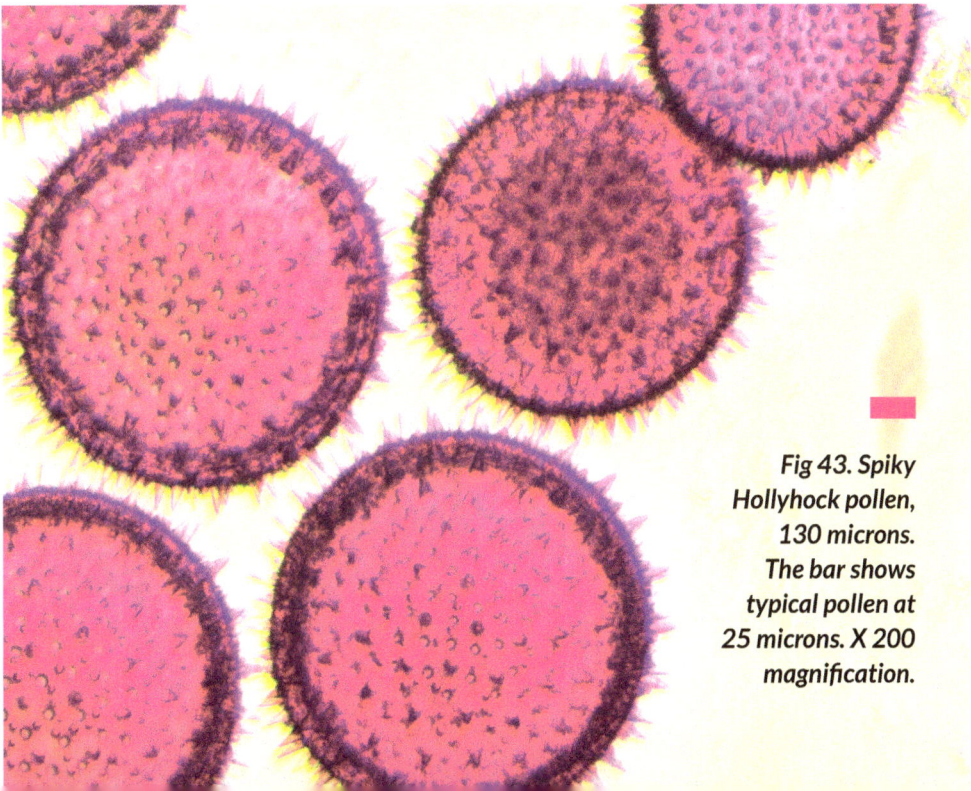

Fig 43. Spiky Hollyhock pollen, 130 microns. The bar shows typical pollen at 25 microns. X 200 magnification.

There are oligolectic bees, a term in pollination ecology to refer to bees that special-ise in collecting pollen from genera/families, or even a single genus of flower, such as those that produce spiny pollen; these have all developed together in their native countries. Birds and moths are also productive in pollination.

That stately cottage plant has a spiky secret brought with it by those plant hunters of old. Luckily, it has managed to fulfil its masculine role and guarantees its survival.

All pictures by Catherine and Graham Kingham.

Pollen showers

Many of the world's most important crop plants are wind-pollinated, this process is called *anemophily*, and these include Wheat, Rice, Corn, Rye, Barley, and Oats. Many economically important trees are also wind-pollinated. These include Pines, Spruces, Firs and many hardwood trees, including several species cultivated for nut produc-tion. The Hazel '*Corylus*' is a good example as it is one of the first trees to send out its pollen; it is used by the honeybee for protein, but unfortunately for humans, it can cause hay fever!

Wind-pollinated plants do not invest in resources that attract pollinating organisms, such as showy flowers, nectar, and scent. Instead, they produce larger quantities of light, dry pollen from small, plain flowers that can be carried on the wind. Female struc-tures on wind-pollinated plants are adapted to capture the passing pollen from the air, but the majority of the pollen goes to waste.

Dehiscence is the splitting at maturity along a built-in line of weakness in a plant struc-ture in order to release its contents, and is common among many plant species. Pollen released to the wind from the Hazel catkins is one such method.

Fig 44. The small insignificant red flower of the Hazel tree can just be seen bottom right, often found as a bud next to the catkins; it is less than 1mm in size. The catkins are in a state of dehiscence, being elongated and open.

Fig 45. If you tap a pine tree when in dehiscence, you can witness the pollen shower.

Sometimes this involves the complete detachment of a part. Structures that do not open in this way are called indehiscent, and rely on other mechanisms such as decay or predation to release the contents.

The other method of pollination is *Zoophily* when a plant relies on animals to transfer pollen.

The plant uses its flower to advertise the presence of resources, including nectar and pollen, and to attract an animal pollinator. When the pollinator visits the flower to collect resources, it deliberately or accidentally picks up pollen on its body. As it goes on to forage at other flowers, an effective pollinator will deliver some of that pollen to a receptive female flower of the same plant species. The honeybee is one of the main pollinators of plants, as it is able to use the ultraviolet light to see shapes displayed by the plants, a bit like a bulls eye advertising its presence.

For the budding microscopist

To collect some Hazel pollen, select a small twig with elongated catkins and place it in a glass with some water, placing a large sheet of paper underneath to catch the pollen dust over the next day or so. A paper bag also placed over the top stops the dust spreading and other pollen in the air contaminating the specimen. The collected pollen is then put into a container with some *isopropyl alcohol* to preserve it until a convenient time and to remove the protective waxy colour coating. The next step is to remove the dirty *alcohol* by decanting carefully and then replacing with clean *alcohol*; this can now be stored permanently in a tight container until needed. For the bee microscopist the Hazel pollen has another role by helping to measuring the size of other pollen grains due to its consistent size at 25µm by comparing the two specimens when combined

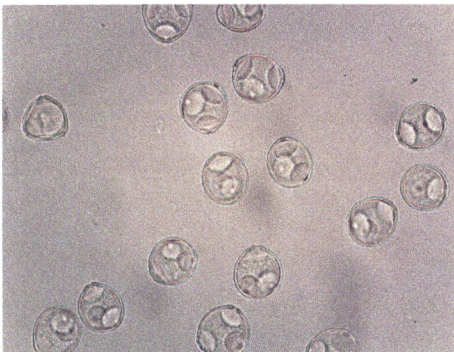

Fig 46. Unstained Hazel pollen, note the triangular shape, smooth surface with 3 apertures, these are the areas where the tube develops from when the pollen meets the stigma, allowing the nucleus to escape to fertilise the egg. X 400 magnification.

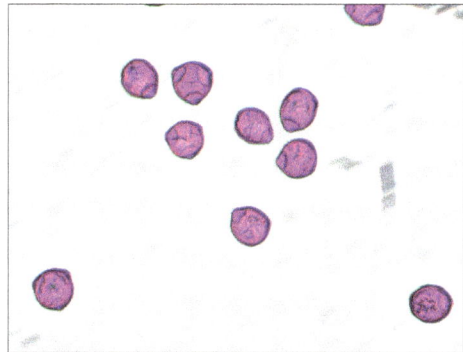

Fig 47. Stained Hazel pollen, the main reason for staining is to darken the subject matter in order to add contrast. By the nature of having to be very thin in order to allow light to be transmitted through it, the specimen is often transparent itself making it hard to see clearly. X 400 magnification.

on the slide. *The expense of buying an eyepiece with a graticule scale and a stage micrometer (a glass slide with millimetre measurements on, these being the standard measuring methods) are in the region of £50.00 plus.*

When preparing a new slide of pollen from another plant a small drop of pollen liquid is added to the slide followed by a small drop of the Hazel mix immediately after the first; this allows the two separate pollen grains to mix. Please use separate rods or you will contaminate your pure samples. The shape, size and structure of pollen vary considerably; each of these special characteristics act as an aid in identifying the plant from which they originate.

A multipurpose plant

The flower of the *Borago officinalis* plant is often to be found in a posh glass of Pimms summer cup alongside some of its cucumber tasting leaves. Our ancestors have used the plant for beer, infused oils, various herbal drinks, and iced teas, in salads and as candied flowers. A whole flower can be frozen in the centre of an individual ice cube for adding to drinks later on in the year, a splash of interest out of season!

This well-established wild plant, that self-seeds, is better known to humans by its popular name, *Borage*. Also known as Starflower, Bee bush, Beebread, and Bugloss, it is a medicinal herb. It has a reputation throughout history of making men and women glad and merry, to comfort the heart, and give courage. The flowers have been used by old painters to colour the Madonna's robes.

This herb is also the highest known plant source of *gamma-linolenic acid*, an *omega 6 fatty acid*, also known as GLA. The seed oil is marketed as a GLA supplement.

It is a hardy annual that will flower in its first year of sowing if sown between March and May, lasting for several months; it is also a favourite of the honeybee, who finds its little vivid blue star shaped blooms irresistible. Bumblebees seem to like it as well.

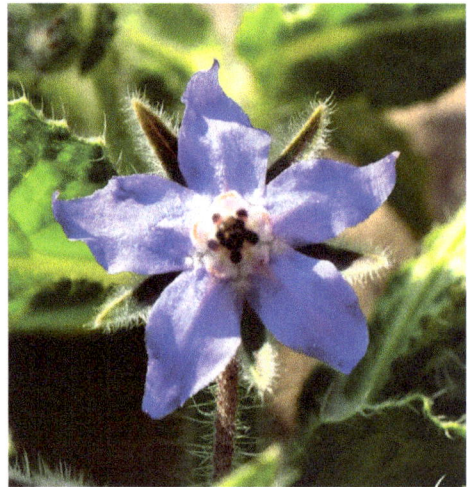

Fig 48. *Waiting for a bee or even an ice cube moment.*

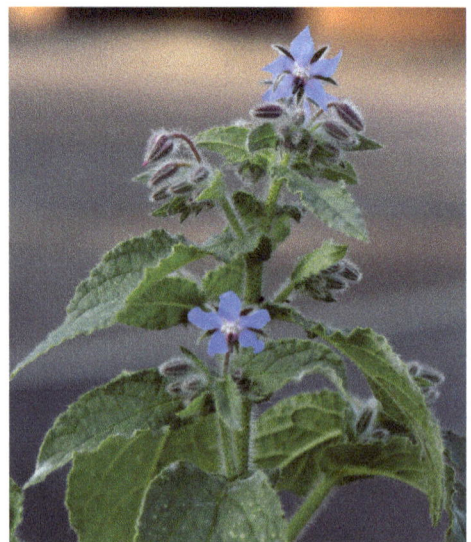

Fig 49. *Borage in bloom, this picture was taken in our garden in July.*

Borage will often grow anywhere and in any type of soil; try planting it in an odd corner to help all sorts of insects and to add colour, or as a companion plant next to broad beans to help attract other pollinators. It is said to stimulate Strawberry growth when planted nearby. Seeds are best sown in full or partial sun under 1centimetre of soil so it is easy to sprinkle a patch with seeds and then cover it with a few handfuls of soil or compost. The plants can easily grow to be 90 centimetres tall and 60 centimetres wide, so give them room to

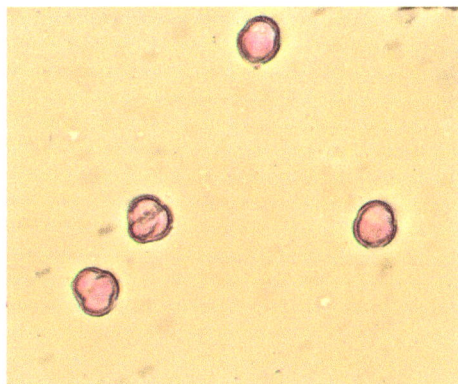

Fig 50. Borage pollen X 400 magnification.

grow, beware of high wind damage, any broken stems can be fed to chickens they will also peck health plants if allowed.

Oregano, the joy of the mountain

Oregano is a culinary and medicinal herb from the Mint, or *Lamiaceae* family. Closely related to *Marjoram*, of which it is the wild equivalent, *Oregano* has a coarser, more robust flavour with a hint of *Thyme*, and a warm aroma.

It has been used in medicine and cooking for thousands of years. The name of the herb comes from the Greek words "*oros*," meaning mountain, and "*ganos*," meaning joy. It typically grows around 50 cm tall and has green leaves around 2 to 3 centimetres in length; it likes free draining soil and can be grown all year round, however our cold winters tend to kill it off. The purple flowers are evident from July to September; the honeybee does not make honey from it but uses it as a pollen source.

Cooking tips

Oregano goes particularly well with Tomatoes, Aubergine, and lamb and is generally added just at the end of cooking, so that it retains its pungency. Start with a small amount, as too much can make the food bitter.

Fig 51. Bushy plant.

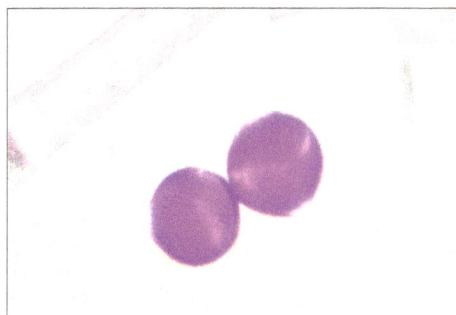

Fig 52. Pollen for last year's crop. X 400 magnification.

One teaspoon of dried *Oregano* is equivalent to one tablespoon of fresh *Oregano*.

The Greeks use it dried sprinkled over their Feta cheese salads. If you visit the hills of Greece, you can see this herb alongside *Thyme* growing wild, perfuming the air on a hot day, magical!

I grow some in the garden and in a pot taking some leaves to dry before it flowers, then I tend to watch the bees visiting the flowers from our kitchen window, it sits next to a large *Lavender* bush where the Bumblebees and honeybees take it in turn to visit each plant.

Mahonia, a bargain plant for all creatures

A stunning winter flowering shrub that producing long woody stems, with evergreen spiky leaves and yellow arching flowers whose scent is reminiscent of Lily-of-the-valley. On a still day, you can catch its perfume in the air. *Mahonia* is related to *Berberis*. The green leaves often take on a rainbow show of colours after a   frosty spell of weather.

The flowers are individually small and globular or bell-like in shapes that are borne on long racemes, (meaning a flower cluster with the separate flowers attached by short equal stalks at equal distances along a central stem. The flowers at the base of the central stem develop first.

They open from the centre of the plant first, and flowerings last for many weeks, and are much used by the honeybee. In the spring honey crop last year, the centre of two super where dark brown, and when a sample was taken and when viewed under the microscope *Mahonia* pollen was present in abundance, evidence of their visits locally! The honey from these supers tasted divine; however, bees cannot make honey from the flowers, as it is only pollen that is available to them. The flowers are often followed by small, plum-coloured fruits, which birds will eat.

Four popular species and hybrids are listed below. Most are hardy in England. Once planted, they need no further care, apart from the occasional removal of dead wood as they are slow growing; however, they can become leggy over time or in shady posi-

Fig 53. Mahonia *showing flowers and coloured foliage.*

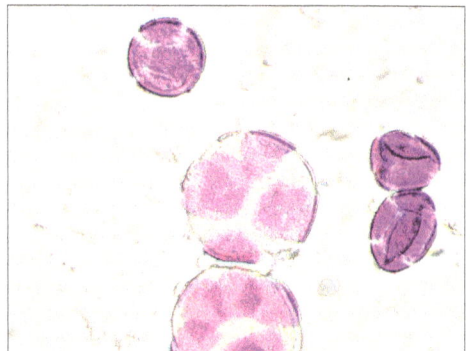

Fig 54. Mahonia *pollen 40 µm X 400 magnification.*

tions, taking up to 3 metres in area and height. Owing to the glossy leaves, they are left alone by pests. Gardeners need to be aware of their sharp prickles.

Mahonia lomariifolia originates from China and Burma, used as one parent of *Mahonia* x media hybrids, such as 'Buckland' and 'Winter Sun'. It reaches 3 metres high with cracked bark.

Mahonia x *media* producing the largest racemes and leaves up to 40 centimetres long.

Mahonia japonica comes from China and is hardy. Takes up an area of about 2 metres square. The leaves emerge from the stems almost horizontally. The winter flowers are yellow and perfumed, held in arching racemes. It also has masses of oval, purple fruit in early summer. This species, actually from China, is a reliable, hardy shrub.

Mahonia aquifolium comes from the western United States and produces blue-black fruits.

A must have plant in our garden because it is low in maintenance, has striking leaves and foliage, and heavenly scent. It provides food for the birds and is an excellent pollen source for the honeybee in winter and early spring.

Tilia and the bees

Whilst enjoying the company of some friends in late July, our host took me outside to show me something of interest. In the midst of his garden an over, powering sweet perfume greeted the nose whilst a constant drone of buzzing met the ears. A tall Small Leaf Lime cordate tree (*Tilia platyphyllos*) stood in front of us; the west facing side was covered in blossom, the flowers were heaving with short tongued bumble bees, buff tailed and red tailed, honeybees and predatory hornets looking for an easy meal; the buzzing noise was audible 10 meters away. A deciduous tree up to 38 meters tall, it has a relatively short trunk and wide low, dense evenly domed crown when growing in the open, but will be tall with unbranched trunk and high crown if crowded amongst other trees. The tree flowers in June and July after the leaves have formed, producing

Fig 55, 56. The yellow flowers grouped together, close up showing pollen.

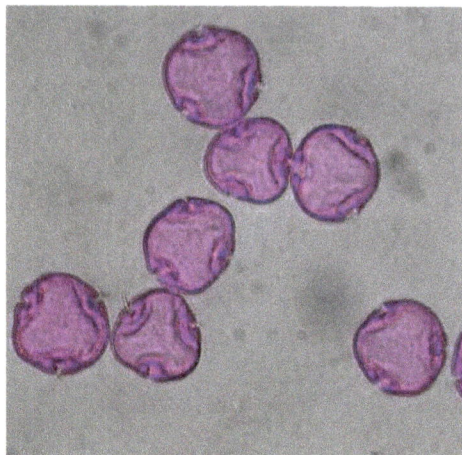

Fig 57, 58. The tall Small Leaf Lime tree. Lime pollen, 35 microns X 400 magnification.

between 4 and 11 yellow, strongly sweet scented *hermaphrodite* flowers, meaning both the male and female reproductive parts are contained within one flower. They produce very sweet nectar & as a result, all bees love this plant, it ranks in the top 10 most important plants for honeybees. Aphids are also attracted to the leaves eating the sap and as a result producing honeydew.

The nectar is produced inside the five star shaped sepals and is held in place by small hairs and surface tension. 20 degrees *Celsius* is deemed the optimum temperature for nectar production; hotter days and wind tend to dry the nectar up. Thin honey is produced being pale to light yellow with a greenish tint, it has a strong biting, minty taste with slight vanilla notes. It is slow to granulate. The Tilia is found throughout Europe and often in mixed lowland forests. It has a variable reputation with regard to regular flowering due to the vagaries of English weather. In France, its flowers are gathered and dried then made into tisanes to drink as a digestive and calming tonic. Woodcarvers prize the close-grained wood, as it is easy to work and does not split.

Polling results in!

Given the choice, honeybees will visit some flowers over others; they will not as a rule visit different species of plants. Bumblebees spend time hopping from flower to flower, as they work their way round the garden.

The honeybees are programmed to visit flowers planted en-masse and one species at a time: this saves energy and they reap the benefits of lots of available pollen to return with to feed the colony, unlike the humble bumblebee who normally only feeds themselves, actually they do return to their nests with pollen for their young. Competition between the two bee species for feeding is partially negated by their differing tongue lengths, allowing differing flowers to be utilised.

The top ten plants for honeybees are listed below, most are abundant in the countryside.

Flower	Nectar or pollen	Flowering times
Borage	Nectar and pollen	April to October
Bramble	Nectar and pollen	May to September
Cherry	Nectar and Pollen	April to May
Dandelion	Nectar and pollen	March to October
Lime	Nectar, some pollen	June to July
Michaelmas daisy	Nectar and pollen	July to October
Orchard fruits	Mainly pollen	April to May
Rosebay willow herb	Nectar and pollen	July to September
White clover	Nectar	June to September
Willow	Mainly pollen	February to May

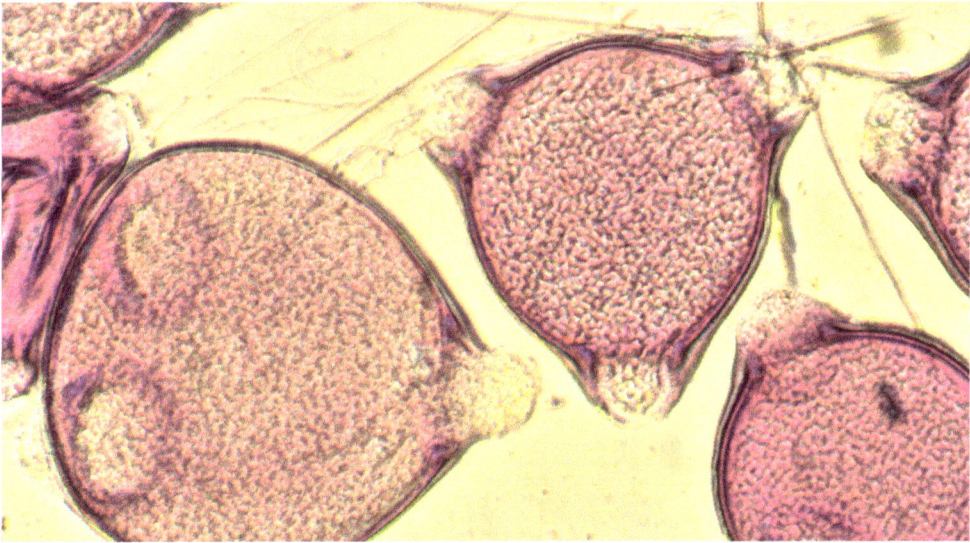

Fig 59. Rosebay willow herb X 400 magnification.

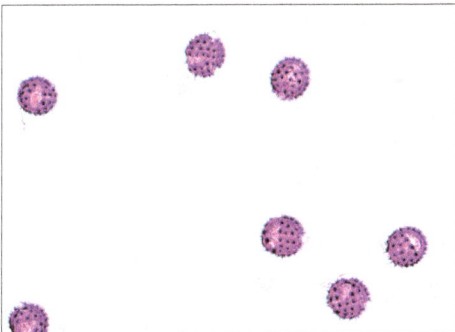

Fig 60. Michaelmas daisy X 400 magnification.

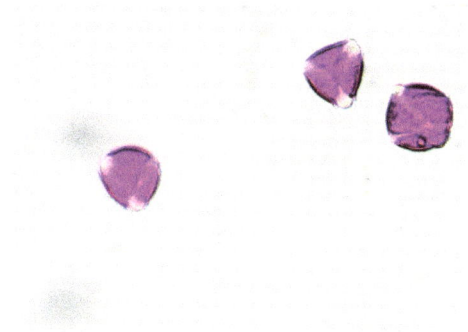

Fig 61. Blackberry X 400 magnification.

Many other flowers both wild and cultivated attract the bees also; please remember to plant them in groups if you want honeybees in your garden. There are a couple of plants that the honeybees do not like Rhododendrons being one; they contain a toxin called *Grayanotoxin* which is poisonous to them but not to the bumblebee. The second reported plant is the native Lime tree mentioned above; in prolonged dry years the honeybees can be found paralysed beneath the tree, along with some bumblebees, this happened locally in July this year in my neigbour's garden. The explanation seems to be that the trees have run out of nectar due to drought conditions and the bees are starving and just too exhausted to fly. Rescue is possible - apparently, you feed them fresh nectar.

This information was taken from the excellent book Plants for Bees. A Guide to Plants that Benefit the Bees of the British Isles. By kind permission of IBRA. The second edition also includes pollen slides for some plants.

Honey pollen 2018 (taken in early June and July)

All of the grains have come from five different hives; all have shared the same pollen load except the Dandelion, which there was only one grain in all 18 slides made. I am unable to conclude the honey source as there was so little pollen in sample.

However, I have made a third set of slides with the sludge! These did have more pollen in and I did not find any Sweet Chestnut pollen. All sample no matter from the source seemed to have the same pollen types in which showed the typical flowers of our area. Major pollen was Blackberry, followed by *bracsica* type.....Red Campion or Oil seed rape, some grass with a few odd grains, mainly Hogweed type. Blackberry honey is of medium colour and course flavour. This just might be the darker honey as we had a mass of flowers this year.

I made several slides from my own honey hoping to condense down the pollen, but I have the same results.

In order to say that honey is one type it must have a majority of that plant pollen present, not all plants produce nectar for honey productions, just pollen. Some pollen is over represented in some samples due to the amount the flowers produce in comparison to others. X 400 magnification on all slides.

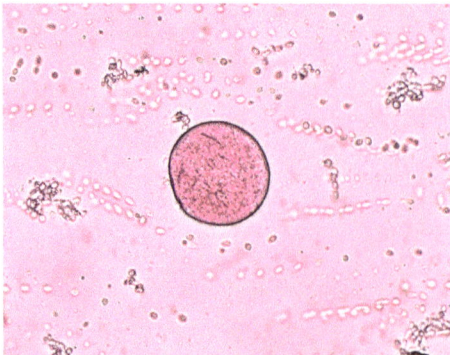

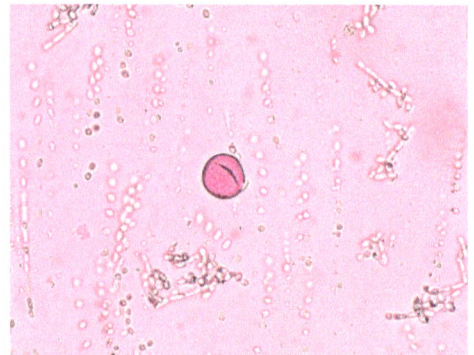

Fig 62, 63. Bracsica - Red Campion. 30 microns. Common

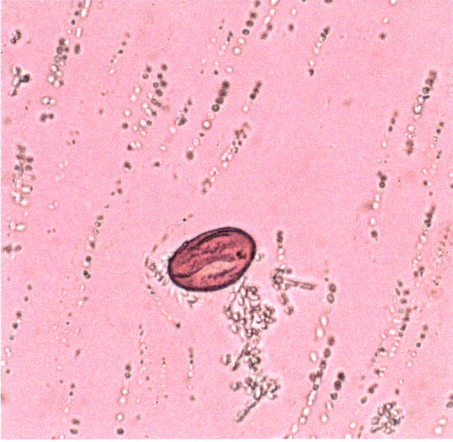

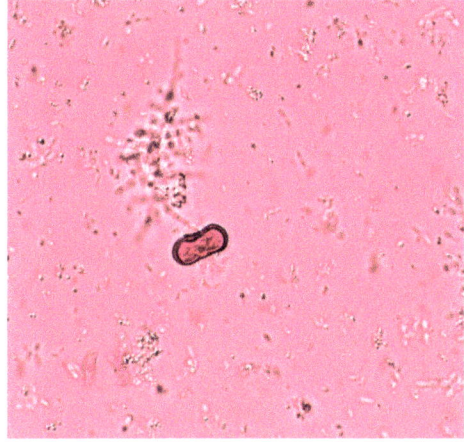

Fig 64, 65. Unknown - not hydrated. Grass possibly. Common / Hogweed Common.

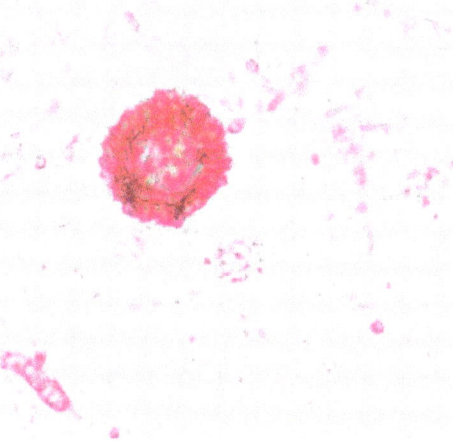

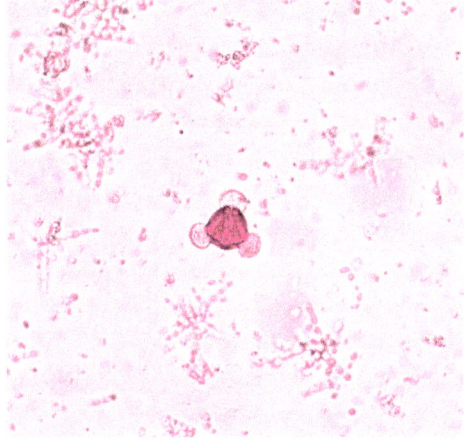

Fig 66, 67. Dandelion. Dark honey 25 microns. Pollen breaking open / Cytoplasm escaping.

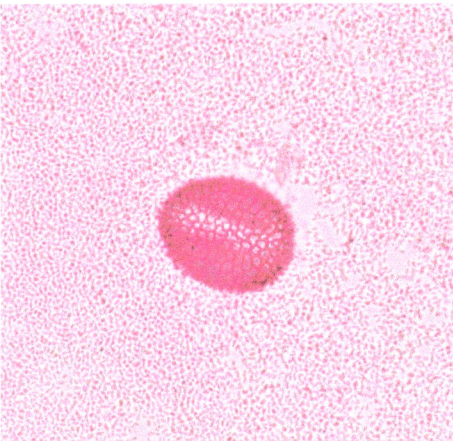

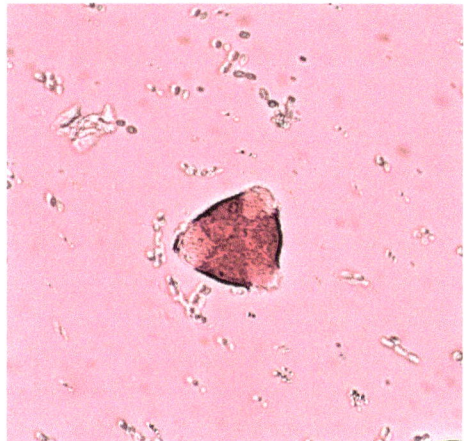

Fig 68, 69. Lily 75 micron. Dark honey only. Blackberry / Common and abundant.

Honey and Pollen Coefficients

This article sets out the basic history of pollen coefficient in honey and the means and methods by which scientists calculate their values; beekeepers who are undertaking the microscopist exam might find it of interest. The BBKA syllabus requires you to 'take into account the over and under representation of pollen in a multifloral honey'.

One of the goals of *melissopalynology*, the study of pollen grains in honey, is to determine the floral sources utilized by honeybees in the production of honey. Because some types of commercial honey are preferred over others, they are in high demand and command much higher prices, think of heather or Manuka. Verification of these premium types of honey is often difficult because many of them come from plant sources that either are weak pollen producers or have pollen that is under represented in honey. In an effort to verify these premium honey types, researchers developed various methods for correcting the pollen data. These methods produce what are known as pollen coefficient (PC) values.

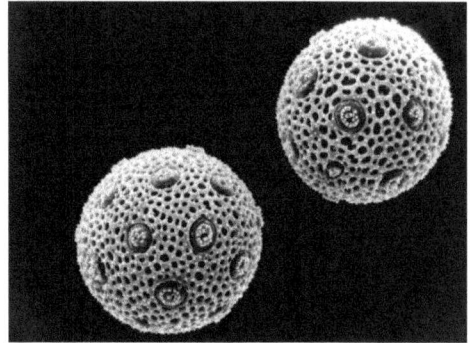

Fig 70. Red Campion pollen, Electron scanning microscope by kind permission of Robert Litchfield, for more spectacular images please visit his web site. www.rowbo.info.

Precision in interpreting pollen data has always been a primary goal of *palynologists*, the general study of pollen grains, for example, when using pollen counts to reconstruct current or past vegetations or determine the nectar sources of a honey sample, the types and percentages of recovered pollen are rarely considered an accurate one-to-one correlation with the floral types they represent. Since the development of pollen analytical techniques during the early 1900s, many advances and improvements have been made in sample collection techniques, in laboratory methods for pollen extraction from matrix materials, and in microscope optics. Nevertheless, once the pollen analyses are completed and the data tabulated, the palynologist must rely on other factors before assigning meaningful interpretations to the data.

First, pollen distribution is initially affected by one or more innate characteristics such as pollen mass, pollen morphology, pollen production and the method of dispersal.

Second, methods of pollen dispersal vary. Some pollen types are released from their *anthers* into the air, named *anemophilous*, become airborne, and are dispersed various distances from their source. The distance of dispersal depends on many factors including air temperature, humidity, pollen sinking speed, changes in surface topography, and the force and direction of the prevailing winds.

Other pollen types are *zoophilous* and rely upon insects, bats, birds, or other small mammals to disperse their pollen.

The methods of collection, transportation, and storage by these insect and animal pollinating vectors play important roles in any subsequent interpretation of the importance of the types and abundance of collected pollen. Finally, once pollen is stored by insects or deposited on the ground as part of the pollen 'rain', its rate of deterioration will be affected by microbial activity, cycles of wetting and drying, pH, chemical oxidation, and mechanical breakdown of the *cellulose* and *sporopollenin* of the pollen outer wall.

There are a number of major variables that affect the accurate determination of where a honey sample was produced and the type of nectar sources that were used to produce the honey.

First, we have learned that field identification of nectar sources used by bees in the production of honey is most often incorrect. After examining 1,100 honey samples from hives in the United States they discovered that more than 60% of those identifications made by the beekeeper or honey producer are incorrect as to the purported nectar sources.

Second, experimental data revealed that honeybees are able to remove a vast amount of pollen from the nectar during their return flight to the hive. In addition, tests reveal that not all honeybees are equally efficient in removing this pollen, as the size and shape of a pollen grain also determines how efficiently honeybees can remove these pollen types.

Third, a growing number of beekeepers and honey producers partially or completely filter their honey before selling it.

Fourth, they have examined a number of standard processing techniques currently used to extract pollen from honey and have found flaws in each method.

Finally, even when honey samples are correctly processed and their pollen contents are carefully noted, the resulting relative pollen data may not provide an accurate view of the primary nectar sources used to produce the honey.

The challenge is to understand the relationship between the pollen sources and the recovered pollen data. In the study of honey, the focus has been on the correct assignment of floral sources and the determination of unifloral honey types.

During the early 1940s, two scientists, Todd and Vansell, working in California examined the relationship between the pollen in the floral sources utilized by honeybees and the importance and recovery of those same pollen types in honey. Their research began when they discovered that bee colonies survived, but would not reproduce when fed only sugar syrup. Once pollen was added to the syrup, the bees began egg laying within twelve hours. A pollen coefficients laboratory was located there and they could get assistance from botanical experts at the University of California at Berkeley. Their research began by collecting and examining over 2,600 individual samples of nectar. They had three major goals.

The first goal was to determine the number of pollen grains one should expect to find in one cubic centimetre of nectar from various plant species.

The second goal was to determine if the number of pollen grains naturally occurring in nectar samples matched the number of grains found in the honey stomachs of the bees that foraged on the same nectar types.

The third goal was to discover how efficiently honeybees removed pollen from the nectars they collected. Although Todd and Vansell did not propose a table of statistical 'R-values' these are the values given to pollen in the wild, to compensate for the over or under representation of pollen types in honey, their data showed that not all plant sources contribute pollen equally to nectar and honey.

They effectively demonstrated that there is not a 1: 1 relationship between a honeybee's use of a plant's nectar and the percentages of pollen contributed by that nectar source to the produced honey. Their research became the foundation for the later development of pollen coefficient values in *melissopalynology*.

Pollen is a honeybee's only source of *proteins, fatty substances, minerals,* and *vitamins*. It is essential for the growth of *larvae* and young adult bees. Honeybees remove pollen from an *anther* by using their *tongue* and *mandibles*. While crawling over flowers, pollen adheres to their 'hairy' legs and body. These are specialized *plumose* hairs. The honeybee combs pollen from these specialized hairs covering her head, body, and forward appendages, mixes it with nectar from her mouth, and transfers it to the *corbiculae*, or 'pollen basket', on her posterior pair of legs.

When loaded with pollen, she will return to her hive. Once at the hive, workers pack the pollen into special comb cells located in the central portion of the hive surrounding the brood area. To prevent bacterial growth and delay pollen germination, a *phytocidal acid* is added to the pollen as it is packed into the comb.

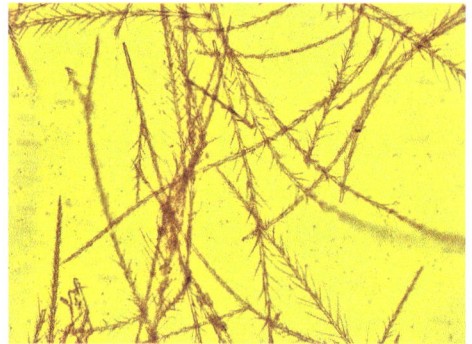

Fig 71. Plumose hairs X 100 magnification.

Other *enzymes* produced by worker bees are also added to prevent *anaerobic metabolism* and fermentation; there by enhancing the longevity of the stored pollen. Once completely processed for storage, the pollen comb referred to as 'bee bread' and is ready for later consumption by the bees. The *protein* source needed for rearing one worker bee from *larval* to adult stage requires approximately 120 to 145 mg of pollen. An average bee colony will collect about 20 to 55 kilos of pollen a year.

In most cases, the primary foraging sources for pollen are the various insect pollinated, *entomophilous* plants honeybees visit for nectar. However, honeybees will also visit a number of species of *anemophilous* plants (wind pollinated) to collect pollen, such as *Salix, willow*. Many species of grasses are important pollen sources for foraging honeybees as pollen can be incorporated into honey in a number of ways. When a honeybee lands on a flower in search of nectar, some of the flower's pollen is dislodged and falls into the nectar that is sucked up by the bee and stored in her stomach. At the same time, other pollen grains can become attached to the 'hairs', legs, *antenna* and even the eyes of visiting bees. Later, some of the pollen that was sucked into her stomach will be regurgitated with the collected nectar and deposited into open comb cells of the hive. While still in the hive, that same honeybee may groom her body in an effort to remove the entangled pollen on her body. During that, process pollen can fall directly into open comb cells or onto areas of the hive where other bees may track it into regions of the hive where unripe honey is still exposed. Airborne pollen is another potential source of pollen in honey. Airborne pollen produced by *anemophilous* plants not usually visited by honeybees can enter a hive on wind currents. These *anemophilous* pollen grains are usually few in number, when compared to the pollen carried into the hive by worker bees; nevertheless, those pollen types regularly enter a hive on air currents and can settle out in areas where open comb cells are being filled with nectar. Sometimes airborne pollen can also be deposited into ripened honey when it is being removed by the beekeeper. The pollen 'rain' for various regions consists mainly of airborne pollen, and is important in forensics, archaeology, and ecology to identify a specific geographic region. As international data, *anemophilous* pollen types are not as useful in *melissopalynology* because they generally form only a minor fraction of the total pollen spectra found in honey.

Pollen is an essential tool in the analyses of honey. The *taxa* of pollen indicate the floral sources utilized by bees to produce honey. As a result, pollen frequency is used to reveal and label a honey sample as to the major and minor plant foraging sources that were used by the honeybees. Even non-premium grades of honey often need to be examined for legal reasons because they must be correctly labelled as to type before being marketed. Only by, identifying and quantifying the pollen in honey can the full range of plant taxa be identified

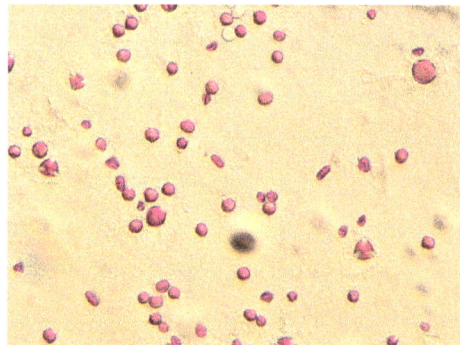

Fig 72. Mixed pollen grains from last year's Devon honey X 100 magnification.

and the honey's actual foraging resources are correctly labelled. Another reason that pollen analyses of honey are often required is to determine the honey's geographical origin. The combination of *anemophilous* and *entomophilous taxa* found in a honey sample will often produce a pollen spectrum that is unique for a specific geographical region. Because of trade agreements, import tariffs, and legal trade restrictions, most of the leading honey producing nations of the world require accurate labelling of honey before it can be sold.

Establishing Pollen Coefficient Values

Early *melissopalynologists*, Whitcomb and Wilson, were studying dysentery in honeybees when they noticed that the bee faeces were filled with pollen grains. They determined that these pollen grains had been sucked into the bee's honey stomach along with nectar during foraging activities. They also noticed that once nectar enters a bee's honey stomach it is filtered. Within about ten minutes, this filtering process removes most of the pollen in the nectar and leaves mostly pure nectar in the honey stomach. The ability of a bee to filter nectar in her honey stomach is important because it is the primary way of removing unwanted debris from nectar, such as pollen and fungal spores, which might germinate and spoil the gathered nectar as it is converted into honey.

The honeybees filtering process is rapid and effective. The bee sucks nectar into a slender tube that ends in the bee's *abdomen* where it becomes an enlarged thin walled sac called the honey *stomach*. This honey *stomach* is greatly distensible and can expand to hold large amounts of nectar. Once in the honey stomach, the nectar passes through the *proventriculus* that serves as a regulatory apparatus, filtering and controlling the entrance of food into the bee's *stomach*. The anterior end of the *proventriculus*, called the honey stopper, projects into the bee's honey *stomach* like the neck of a bottle. At its anterior end is an x-shaped opening consisting of four, thick, triangular-shaped, muscle-controlled lips.

Nectar in the honey stomach is drawn back and forth into the funnel shaped *proventriculus*. This process filters the nectar and removes debris such as pollen grains and the causative organisms of foul brood. The posterior end of the *proventriculus* extends into the anterior end of the *ventriculus*, the part of the bee's *alimentary* canal (mid gut) where digestion and food absorption occurs. A valve at the bottom of the *proventriculus* prevents the filtered nectar from entering the bee's digestive system, but allows the debris removed from the nectar to pass into the bee's *alimentary* canal and into the intestines where it is first stored then later voided from the *rectum*. When large numbers of bees forage on nectars that are laden with pollen, the rapid

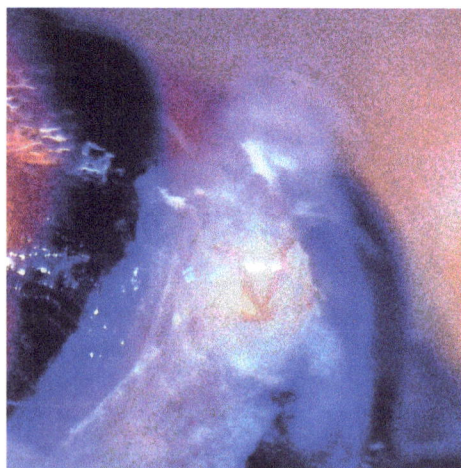

Fig 73. The four orange coloured lips of the proventriculus, honey sac removed X 40 magnification. Picture colour enhanced.

removal of those pollen grains from their honey stomachs and the resulting defecation can appear as 'yellow rain' spots on leaves, cars, paths, washing or buildings.

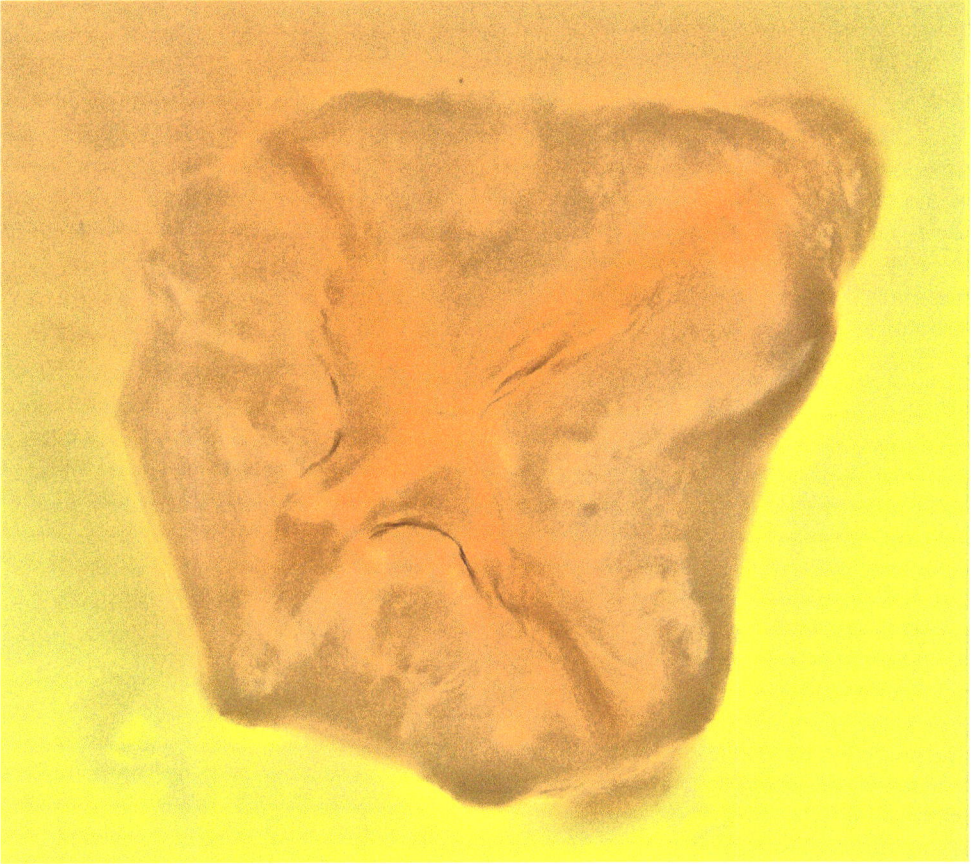

Fig 74. Close up of lips X 200 magnification.

Tireless work

Demianowicz became one of the earliest *melissopalynologists* who worked tirelessly for thirteen years trying to resolve the problem of identifying unifloral honey types based on pollen contents. After examining many honey samples, she realized that the relative pollen counts in honey did not always reflect the primary floral and nectar sources.

To develop the data for each unifloral type, she used caged hives of only 300-400 workers bees and one queen. Fresh, open flowers of a single species were brought to the caged bees several times a day. Under these controlled conditions, Demianowicz believed that the honey produced by each hive was a valid representation of the expected absolute pollen concentration (APC) for the flower species being examined. Based on this research, she developed nineteen different categories of plants ranked based on whether their APC values in honey were under or over represented. Each category was assigned an 'average number' that was determined by averaging the totals of each type in that category. Each of these nineteen categories was called a 'pollen coefficient class.' She was the first to use that term and she believed that the newly established pollen coefficient values could be used as a guide for determining the true unifloral nature of honey samples from any region, regardless of the data implied by the relative pollen counts.

Moar questions the technique used by Louveaux and others for determining unifloral honey types.

First, he points out that there should be exceptions to the Louveaux statement that 45% is the minimal amount for a single pollen type in a unifloral honey. Moar agrees that this might be true for pollen types in honey samples containing between 20,000-100,000 pollen grains per 10g of honey, but should be adjusted for honey samples when concentration values are less than 20,000 pollen grains per 10 g.

Second, he agrees that white clover, *Trifolium repens,* should be considered the base-line for determining coefficient values for other pollen types in honey. However, Moar questions the APC value of 18,000 for *T. repens,* as calculated from the experiment with caged bees by Demianowicz. Instead, Moar uses an APC value of 23,116 grains per 10 g of honey for *T. repens.* Perhaps the difference between the two APC values for *T. repens* (Moar vs. Demianowicz) results from the different ways the bees were allowed to collect honey, free verses caged or different methods each used to calculate the APC value.

Third, Moar proposes new ways to establish baseline APC ratios for various pollen taxa in honey samples produced by honeybees that are allowed to forage freely, when observations reveal that they are visiting mainly the flowers of one plant type.

In summary

Moar presents an example of how to calculate APC values for various plant taxa in New Zealand. He then uses those data to produce pollen coefficient values for those pollen types. He notes, for example, that New Zealand *thyme,* honey is considered a premium commercial type, but based on Louveaux unifloral measuring standards; none of the New Zealand *thyme* honeys would be accepted as unifloral.

Moar notes that in the New Zealand honey samples, *thyme* pollen rarely reaches the minimally needed 45% because its pollen is under represented. Other *melissopalynologists* including Demianowicz and Sawyer also report that *Thyme* and many other pollen types in the mint family are underrepresented in honey samples. In view of those data, it is noteworthy that Tsigouri and Passaloglou-Katrali reported that they found relative percentages of *Thyme* as high as 80% in some unifloral honey samples from Greece where it grows wild in the hills, perfuming the air on a hot day.

In Moar's study, he examined honey samples that were produced by beehives located in or close to abundant fields of blooming *thyme.* The honey produced in those hives was then examined to ensure that it had the traditional colour and taste of *Thyme* honey. Four separate honeycomb samples from various hives that fit these criteria were then processed to extract the pollen. The average relative pollen percentage of *Thyme* in the four samples was four pollen samples 42%. He then compared his pollen counts of *thyme* against the number of tracer spores counted in each sample.

The result was an average APC value of 5,415/10 g of honey for *Thyme* pollen. Moar reasoned that because the relative pollen percentage of *Thyme* was less than the standard 45% needed for unifloral classification, the *Thyme* pollen's APC 5,415, could be adjusted by multiplying it times .45 (the needed percentage). Next, to calculate the 'corrected APC' value for *Thyme* at the 45% level, Moar divided the resulting sum by .42, which is the average relative pollen frequency (42%) of *Thyme* pollen in the four samples. These calculations increased the number of 'expected' *Thyme* pollen in 10g of honey to 5,801, which Moar considered the appropriate 'corrected APC' for *Thyme* pollen at the internationally accepted unifloral level of 45%.

Next, Moar notes that because *thyme* pollen is considered under-represented in honey i.e., any *taxon* with an APC under 20,000 per 10 g of honey. New calculations were necessary to determine what the minimum percentage of *thyme* pollen in a honey sample should be in order to classify that sample as being a unifloral *Thyme* honey. Because the APC of white Clover is considered the baseline standard for honey studies. Moar uses the ratio of *Thyme*'s actual APC of 5,801 against the baseline APC of white clover 23,116 in order to determine the minimal percentage of *thyme* needed in a unifloral honey. Moar proposes that by using his technique any *melissopalynologists* can determine the minimal amount of pollen needed for unifloral classification of any under-represented floral source in any region.

The English contribution

Rex Sawyer, one of the foremost early *melissopalynologists* in the United Kingdom, was a pioneer in the study of pollen analysis. Sawyer's contribution was to use pollen coefficient tables to determine the actual or expected nectar composition of each plant *taxon* in a honey sample; the relative pollen spectrum must first be calculated. Then the relative percentage of each pollen type is divided by its PC value. The resulting value for each pollen type is the *taxon*'s 'relative quantity RQ.' Finally, each RQ value is divided by the sum of all RQ values to determine what percentage of the honey's nectar was actually derived from each plant type.

Sawyer has left a legacy of books behind, about pollen identification for beekeepers, both of which are readily available in paperback; a must read for budding microscopist.

My grateful thanks go to the authors Vaughn M Bryan Jr and Gretchen Jones for allowing me to summarize their article, the full version can be found on the web. Also to Master Bee keeper Chris Utting for his constant advice and help.

http://entnemdept.ufl.edu/honeybee/extension/Honey%20Show%20and%20Judging/Bryant%20%20Jones%20%282001%29%20The%20R-Values%20of%20Honey%20-%20Coefficients.pdf

Pollen abound

North Devon apiary took part in the Citizen Scientist Investigation, Pollen project collection this year; they had to collect pollen samples from three different hives for 1 day only, every 3 weeks from April until October using the pollen traps supplied.

CSI pollen is a worldwide study of the diversity of pollen sources collected by honeybees, organised by the COLOSS honeybee research association; http://coloss.org/.

It is known that the pollen of certain plant species may be deficient in certain essential nutrients and lead to stress in honeybees. In order to investigate regional patterns of pollen diversity, the project sought the help of the local volunteered Citizen Scientists.

The aims of the study are:

- To identify regions with high or low pollen diversity.
- To identify seasonal changes in pollen diversity available to honeybees.
- To compare habitats e.g. city vs. agricultural.
- To enable beekeepers to simply compare and judge pollen diversity of different apiaries.

Two volunteers, into a honey jar, collected the pollen and the coloured pollen loads were then separated into different colour samples the number of different colours present where counted, colour blindness being taken into account!

The pollen was then classified by the number of pollen loads per colour as follows:

ABUNDANT colours in the sample >20 pollen pellets.

RARE colours in the sample 3-20 pollen pellets.

VERY RARE colours in the sample 1-2 pollen pellets.

The findings are initially added to the database via the internet then 20 g of pollen were posted off to the University of Cardiff where a sample of each will be analyzed and identified during 2016. Any spare pollen was given away for human consumption.

The group was astounded to collect over a pound of pollen from an individual hive on some days.

North Devon apiary also decided to make their own pollen slides from each of the 59 pollen samples and to note the time of year collected, the colour and the plant identification using a microscope, several reference books and members with local plant knowledge; a listing of some of the actual local food plants is set out below.

C. S. I. Pollen

Citizen Scientist Investigation on pollen diversity forage available to honey bees

Fig 75.

It has become evident that a vast variety of pollen is taken by the honeybee as a food source and although they will collect a large amount of one species when it is, abundant they benefit from a mixed diet.

Pollen Colour	Spring	Summer	Autumn	Total samples
White	Blackberry	Meadow sweet	Himalayan balsam	7
Cream	Cow parsley			7
Yellow	Crocus	Sweet chestnut		11
Dark yellow	Knapweed	Bindweed	Heather	2
Orange	Dandelion	Ragwort	Ivy	8
Dark orange	Snowdrop	Broom		3
Pink	Red dead nettle	Horse chestnut		1
Brown	Gorse	Clover		4
Black	Oriental poppy	Broad bean	Marsh mellow	2
Grey	Spanish blue bell	Rose bay willow herb	Astilbe	3
Olive	Rowan	Buddleia		2
Green	Holly	Pine		4
Blue	Bluebell	Viper bugloss		2

The most popular colours are white, yellow and orange; however these pollen loads only reflect the abundance of local plants in bloom and depend on how long a period that they remain in flower - as an example dandelion can be found in flower over 5 months or more.

The colour of the pollen sample will vary from the same plant depending on the weather, time of day collected, the age of the pollen and the added complication of the honeybee's saliva, which is used to bind the pollen together.

The samples were examined under natural light from a window using a white sheet of paper as background.

We eagerly await the results if only to check our own findings and to better understand our bee's actual needs.

Fig 76. Pollen trap in situ.

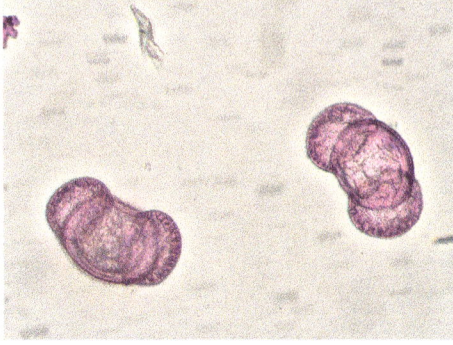
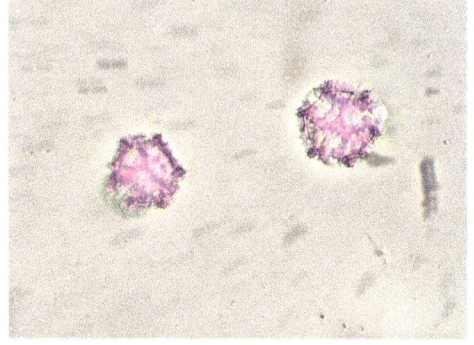

Fig 77, 78. Pollen samples. Pine / Dandelion X 400 magnification.

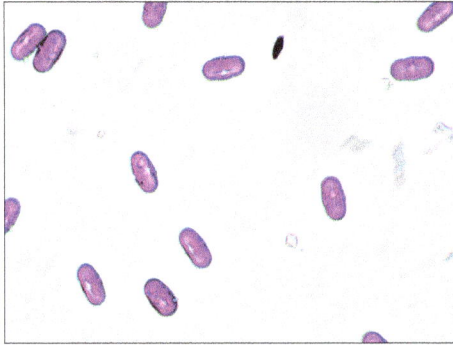
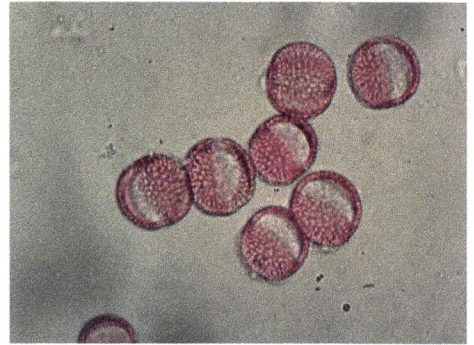

Fig 79, 80. Hogweed / Privet X 400 magnification.

Daffodil camouflage

It is obviously not fooled by the decoy. Remember that consideration of water must be taken into account when siting a hive.

Fig 81. A thirsty bee taking advantage of a pool of water outside of our conservatory during the HOT first, May bank holiday when the temperature reached 24 degrees Centigrade! Picture by Catherine Kingham.

Ivy, food for all

There are two native subspecies of ivy in the British Isles: *Hedera helix* species *Helix*, and *Hedera helix* species *Hibernica*. The subspecies *Hibernica* does not climb but spreads across the ground. There are also many cultivated varieties of ivy, with differing leaves, which are variable in size, colour, number and depth of lobes. The leaves are often variegated green with white, cream or yellow.

Ivy got its name from the Old English word *ifig* and is used as a name for females in England.

Ivy is an evergreen, woody climber; with specialised hairs on their stems that help them stick to surfaces as it climbs. It can grow to a height of up to 30m. It has two different forms - juvenile and mature. Only the mature form of ivy produces flowers and fruits; they are yellowish green and appear in small clusters known as umbels. It flowers from September to November and fruits are black and globular in clusters & ripen in November to January. The leaves are dark green and glossy with pale veins. Leaves of juvenile forms have 3-5 lobes and a pale underside. On mature forms, leaves are oval or heart shaped without lobes and can be self-supporting.

The pollen is dull yellow in colour; the nectar is very concentrated which helps the bees when trying to remove the excess water when making honey. The honey is white to greenish in colour with a pleasantly aromatic flavour, although it is not to every ones taste. It sets soft and will often be left by the bees in late winter, as they seem to prefer it as nectar.

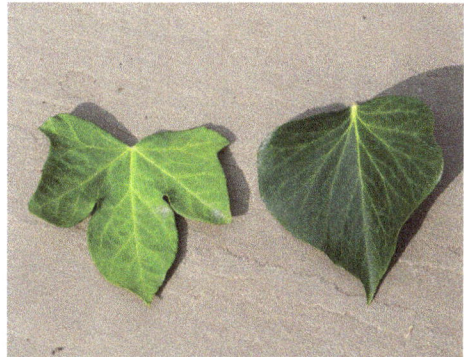

Fig 82. Immature (left) and mature (right) leaf.

Ivy grows well throughout the UK and can be found in many habitats including woodland, scrub, and wasteland and on isolated trees. It is tolerant of shade and survives in all but the most dry, waterlogged or acidic soils. Nectar, pollen and berries of ivy are an essential food source for insects including honeybees and birds during autumn and winter when food is scarce. It also provides shelter for insects, birds, bats and other small mammals. The high fat content of the berries is a nutritious food resource for birds and they are eaten by a range of species including thrushes, blackcaps, woodpigeons and blackbirds. All parts of the plant are toxic to humans. Ivy uses trees

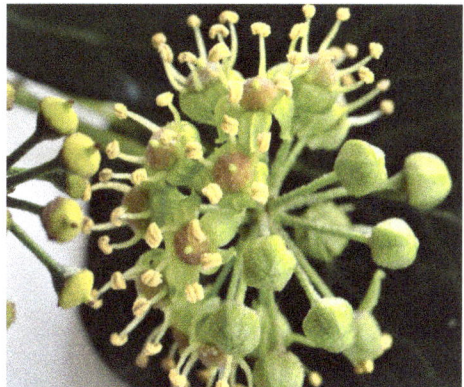

Fig 83. Close up of flowers.

and walls for support, allowing it to reach upwards to better levels of sunlight. It is not a parasitic plant, has a separate root system in the soil, and so absorbs its own nutrients and water as needed. Does ivy kill trees? Only if it becomes too prolific and over-bearing, blocking out light and too heavy for the tree to support.

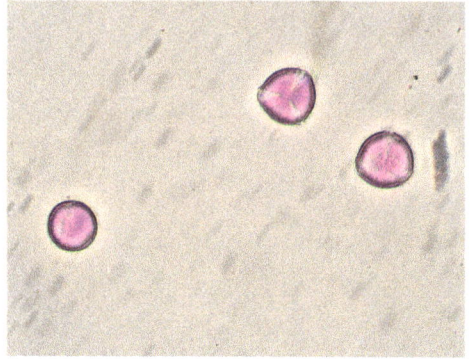

Fig 84. Ivy pollen. 30 microns.
X 400 magnification.

Out & about; challenges

Can you guess what it is?

Fig 85. An old shedded skin of a wax moth larva. X 40 magnification.

I collected the rubbish off the hive floor on second of February and decided to look at it under the microscope at times 10 magnification. There were many loose cappings and a few stray wax plates, which indicated that there might well be a few eggs being laid, or at least some preparation for them. This was of some comfort as this hive had a kilo of candy added the week before, having used up its winter stores. Interestingly my other hive is a poly type and they still have some stores left, for now!

Fig 86. Small Wax moth larva.
X 10 magnification.

Apart from our old enemy the *Varroa* mite, which were evident in very low numbers, the wiggly *larvae* shown was also present in low numbers, but what is it? I had noted the odd larger one in the corner of the hive late last year, but with no damage to the comb. The next inspection on a warm day will reveal all. My conclusion, *Wax moth*.

Observations

Midsummer 2016 I re queened my hive and by autumn the hive was well stocked with worker bees, a very fecund queen me thinks!

After feeding sugar syrup to up the stores, they went into winter well prepared; I hoisted the hive ever 2 weeks using a spring balance on either side, to monitor the load and inserted the *Varroa* board in February 2017 to inspect the crumbs.

So far so good, the odd mite and signs of capping being opened up.

Come March I noted that the hive was light and fed some candy, which they did not consume at all over the next few weeks. On warm days they were flying out, pollen

Fig 87, 88. Propolis and two types of pollen, orange and yellow, Wax moth poo! Wax plate and capping crumbs X 10 magnification.

from dandelions was evident on the floor, together with propolis, sugar crystals, old cappings, the odd dead *Varroa* mite, bee legs and some wax plates.

The next visual inspection on a warm day when many bees where flying about, these viewing take place often as my hives are kept at the bottom of our garden. I intended to open them up for the first internal inspection in late March when it was above 15 degrees *Centigrade*. This took place two weeks later and my first concern was the lack of flying bees from one of the hives, followed by a look at the bottom inserted *Varroa* board, a few lines of detritus and 20-day-old, eggs were in evidence, a laying worker was suspected.

Sure enough, no queen could be found, there were lots of eggs about, 2-3 in each cell; there was also capped brood and uncapped *larvae*, which would indicate the queen was alive recently; no queen cells to be found.

As the hive by now was reducing in size I decided to let nature take its course, the *larvae* was capped off and new bees emerged, daily activity could be seen outside with lots of pollen being delivered.

The final evidence can be seen below - mass cappings and sugar crystal on the floor, dropped down as the final bees emerged, then consuming the remaining stores.

Fig 89. A mass of wax cappings and sugar crystal.

The last thing of note was there were 2 distant honey bee strains in the hive, those with brown-yellow *abdomens* and an equal amount with black *abdomens*, who had the queen mated with I wonder?

Noseama, microsporidian or fungus?

Scientists are at loggerheads when trying to classify the Noseama bug, the count is still out!

This disease is common amongst bees; it tends to be more prevalent in the spring and healthy colonies seem to cope with it well. There is no recommended treatment in the UK other than a comb change to remove the spores. The bug settles in the bee's *stomach* and produces spores, which then give the bees *diarrhoea*; this is cleaned up by the young house bees making them more at risk than the rest of the hive. Spare a thought for the humble bumblebee; Noseama affects them also, but the queen and workers seem to have some tolerance; it is the male who suffers. The drones' gut becomes so distended making them unable to fly and because of this, they cannot mate; this then leads to the collapse of the Bumblebee colony.

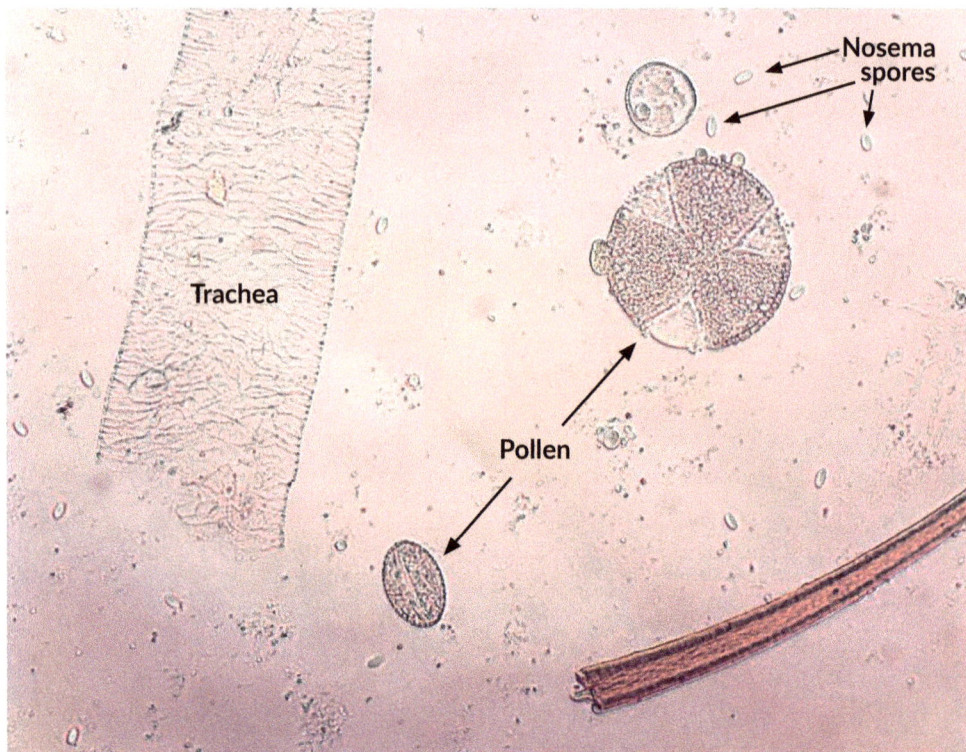

Fig 90. A microscopist view of a Nosema slide X 400 magnification.

Two species of *Nosema* have been identified in the UK, *Nosema apis* and *Nosema ceranae*. These are both fungal pathogens, which invade digestive cells lining the mid-gut of the bee.

Causes

Nosema occurs when fungal spores are ingested by adult honeybees. Symptoms are more apparent when nutrition is poor; weather conditions are cold and wet and during long periods of confinement, typically spring in the UK.

Nosema may be hard to diagnose, as there are no real symptoms of the disease. *Dysentery* is often associated with *Nosema apis* infection and can be characterized by brown spotting across the hive. It is not caused by the pathogen but as a result of infection. During long periods of confinement, some will defecate in the hive, leading to further infection. Infection by *Nosema apis* is also associated with bees crawling around the hive entrance, sometimes with wings held at odd angles. Some bees will have swollen and greasy abdomens and in severe cases may appear to be trembling. With *Nosema ceranae*, it has been reported that infections are characterised by a reduced number of bees in the colony, increased likelihood of disease and eventual collapse.

Beekeepers can spread *Nosema* by transferring contaminated combs between colonies. The spores can remain viable for up to a year. Bees can also spread the disease by feeding on contaminated food and water, cleaning contaminated combs, robbing contaminated hives or drifting to new hives.

The simplest method to diagnose a *Nosema* infection is by microscopic examination. Both species are virtually identical when viewed using conventional microscopy, but can be distinguished by an expert eye.

Beekeepers can no longer use Fumidil B to treat for *Nosema* (authorisation expired on the 31st December 2011). The best way to deal with *Nosema* is with preventative measures i.e. maintaining strong, well-fed colonies headed by young queens, which are more able to withstand infections. It may help to requeen susceptible colonies with a queen from more tolerant stocks better able to cope with a *Nosema* infection.

Wanted, some better PR! Wasps of the British Isles

During the month of April, I put up an insect trap to try to capture the Asian hornet; none were found but several wasp queens and a hornet were trapped; this gave me an opportunity to look at these insects more closely, including dissecting the abdomens. All pictures are taken from the catch.

The wasp's sting has very small barbs like that of the honeybee, which uses its sting

Fig 91. Wasp stinger.
X 200 magnification.

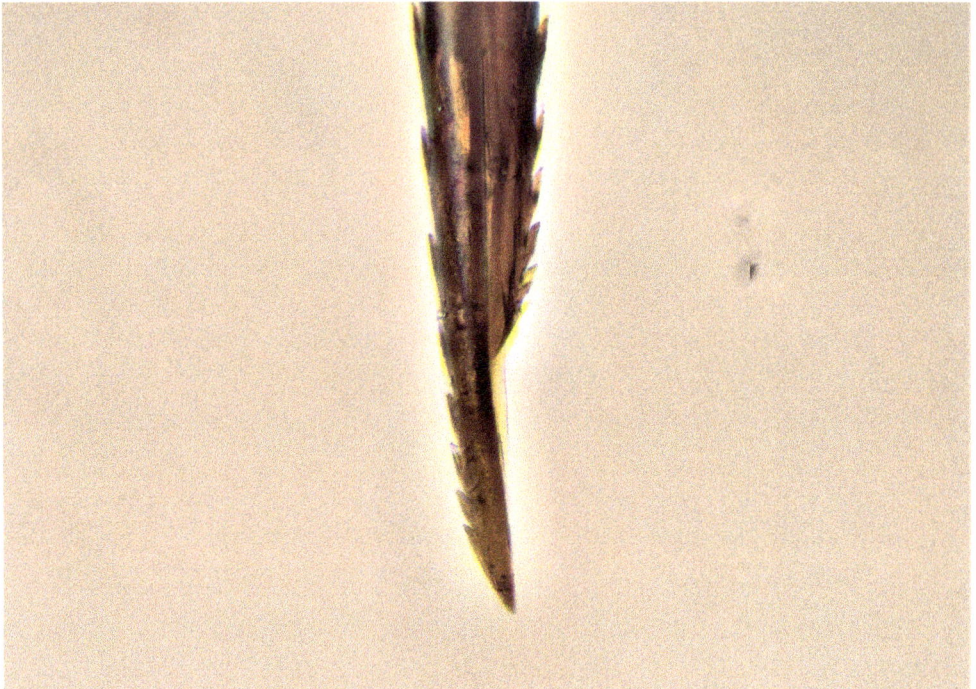

Fig 92. Honeybee stinger showing barbs. X 200 magnification.

purely as a defence; instead, it is used for subduing prey, then returning with the prey to the nest for consumption by the wasp's hungry grubs. This is why the wasp can and often will sting you multiple times without any trauma to itself.

Factors that may serve to influence the wasp's response include - temperature, nest size and nest maturity. Wasps' nests tend to become more of a risk as the wasp matures and the number of workers begins to out-number the *larvae*. This causes a short-age of the sweet sticky treat that the *larvae* provide to the returning worker wasps in exchange for food; it is at this stage in late summer that we see wasps at the lunch table when we eat outside; they are searching for that sugar fix.

If you watch wasps leaving a nest, they will head out in different directions. As you get closer to the nest, you begin to interrupt this flight pattern and if you get too close, you will inevitably get stung or even swarmed by them.

Hymenoptera insects, e.g. ants, bees, and wasps, first appeared in the *Triassic* Period more than 200 million years ago. A large number are extremely social insects and have developed regimented social systems in which members divide into worker, drone, and queen castes. This is referred to as eusocial.

The Egyptians tell of a monarch who died from a wasp sting in 2600 BC, describing the first reported account of a wasp and for that matter anaphylactic shock historically.

The family *Vespidae* is a considerable and diverse group, made up of over 6000 species that are found throughout the world. All UK wasps are all clearly identifiable by their

Fig 93. Wasp on a pear tree.

black and yellow/orange warning decoration designed so this would appear to help predators learn quickly that these insects are not a food source.

Most British social wasp colonies begin in the spring when the hibernating queen emerges in the first mild days of spring. At this time of year, the newly emerged queen is at great risk as she lacks a nest to protect her from late frosts, and until she has fully recovered from her long sleep, is lethargic and unable to defend herself. She will also be one of the few large insects around in early spring so is an obvious target for predators such as birds. Ants are also predators of wasps and given the opportunity will attack them.

Wasps however secrete a substance around the petiole (the stalk that attaches the initial structure of the nest to whatever is supporting it) this acts as a repellent.

Before the queen can begin laying eggs, she first needs to regain her strength by aiding early pollination of plants as she consumes carbohydrate rich nectar and sap; this is why trapping them with a sugar solution works in early spring.

Depending on the preference of the queen, which may vary in different species, a nesting site may be established in all manner of places. Some are subterranean nesting in disused rodent burrows or in naturally occurring hollows in trees. Others will nest in terrestrial nests in structures such as houses, outbuildings, bird boxes, and compost bins. Finally, we have aerial nesters or those that prefer to nest in trees and shrubs or on the sides of structures.

Another reason for nesting in a particular place is believed to be linked with the odour produced by different species of structural timbers, Cedar and Oak. Wasp nests are simply made from whatever material is most abundant and this is invariably chewed up timber, mixed with water and saliva to form wood pulp, essentially paper. This is why social wasps are often referred as the paper wasps. The Wasp can only eat liquid foods as its mouth parts are like the Honeybee, designed to suck. They use their mandibles for cutting up insects and chewing material for making the nest.

The queen wasp will begin building her nest by first establishing a petiole or short supporting spindle on which to mount the first module or layer of hexagonal brood cells. This module is in the shape of a small disk divided into approximately sixteen hexagonal brood cells or chambers. This number can vary considerably.

The next phase for the queen wasp is the laying of a single egg into the base of each new cell. Over the coming three to four weeks, depending on temperature and external conditions, the queen wasp will raise the developing *larvae*, feeding them on a diet rich in insect protein. She does this and her emerging workers will do this as well by finding insects and insect *larvae* and injecting them with *venom*. The *venom* injected disables the prey by paralysis and allows the queen to dissect the prey as required. Caterpillars are often taken to the nest whole, but flying insects have the head, legs, wings and *abdomen* removed as the central unit of the body has the greatest protein concentrations due to containing the muscles responsible for flight and articulation of the legs.

As the *larvae pupate*, the queen is freed to continue nest construction and as the first brood begin to emerge, the queen immediately cleans an empty chamber and lays another egg into it. As the brood grows the petiole is enlarged and yet another, larger horizontal layer or disk is created. Where the space is confined such as in wall cavities, subsequent layers may be created to fit the cavity.

Eventually the queen will only have the job of laying eggs and the nest will continue to mature, with most wasps nest populations being in the region of 3,000 to 10,000. In some species such as the Hornet (*Vespa crabro*), this number will be far less, with only a few hundred individuals.

Fig 94. Comparison of size between a small wasp and a native hornet.

Climate plays a large part in population; in cold weather, this could be four weeks, but in an exceptionally hot summer, this could be only a week. The best conditions are hot with good amounts of rain as it produces an increase in flying insect numbers and ensures the nutritional needs of the nest are met easily.

Later in the summer or as the colony matures, males will develop and leave the nest to mate. Males do not sting, as they do not possess the modified ovipositor or egg laying tube that their mother and sisters possess. At the same time, new queens will also be emerging generously equipped with a fully functional sting. Once mated they will normally go into hibernation.

If conditions are mild, social wasps in the UK will mature more quickly and the emerging queens will create new nests in the same season.

There are nine British wasps plus the hornet who is also a member of the *Vespula* family.

The Common Wasp (*Vespula vulgaris*)

Length up to 20mm long with a black and yellow body but no black spots on its back and has an anchor shape on its face. Likes to nest in hollows in trees or in the ground with up to 10,000 individuals.

This is by far the most commonly encountered British wasp.

Fig 95. Common wasp X 10 magnification.

The German Wasp (*Vespula germanica*)

Length up to 20mm long black and yellow with black spots on its abdomen, the face has three black spots on its face. Likes to nest in hollows in trees or in the ground with up to 10,000 individuals.

The German Wasp joins *V.vulgaris* as one of the two most abundant wasp species in the British Isles. It is wide spread throughout the northern hemisphere.

The Red Wasp (*Vespula rufa*)

This species is up to 18 mm long and found across much of the UK where it prefers rural environments. Unlike other wasps, this species tends to represent less of a nuisance or health risk to humans as it does not scavenge for meat, and the adults feed on nectar.

The Norwegian Wasp (*Dolichovespula norwegica*)

The Norwegian Wasp is more common in the north of the country and is among the dominant species in Scotland. This wasp is also known as the tree wasp and is associated with an aggressive temperament.

The Cuckoo Wasps (*Dolichovespula adultarina* and *austriaca*)

These species are parasites of the red wasp (*Vespula rufa*) and the Saxon wasp (*Dolichovespula saxonica*). The queen enters the nest of the red wasp once the nest is well developed and attempts to kill the defending queen. If successful, she will kill all the existing offspring of their previous queen and lay her own eggs in the nest.

The Tree Wasp (*Dolichovespula sylvestris*)

Dolichovespula sylvestris is an aggressive species not normally encountered in the far south. Its prominent black spot in the centre of its face most easily identifies it.

The Saxon Wasp (*Dolichovespula saxonica*)

18mm long, black and yellow with no black spots on its abdomen, has an irregular black line down its face.

The Saxon Wasp is a relative new comer to the UK, found mainly in the south.

Commonly a nest will be about the size of a grapefruit with around 1,000 – 1,500 individuals however it may grow to well over 20 cm if conditions are conducive.

Fig 96. Saxon wasp X 10 magnification. The Median Wasp (Dolichovespula media).

This aggressive species has a prominent sting that has been measured extending 3mm from the abdomen. The Median Wasp appeared in the UK in the 1980s and has spread across the southern half of the country.

The Hornet, (*Vespa crabro*) a quick visual guide.

Fig 97. The hornet has no black marking on its face. X 10 magnification.

Fig 98. The distinct red bar of the Hornet, which Wasps do not have. X 10 magnification.

Woodlice, opportunists in the hive

Having put my bottom board in to check on the *varroa* drop in December, I noticed that a lot of frass was gathering at the edges but no evidence of its owners. A few days later, during which we had a week of inclement weather, I went out to remove the board, only to find the culprits, over 100 woodlice. It was quite damp at the edges and they seemed to be located there; however, lots more were all over the bottom making the most of the rubbish.

I have never seen any inside the hive and the few bees that had crawled up the front gap did not seem to be bothered by them.

Woodlice may look like insects, but in fact, they are crustaceans and are related to crabs and lobsters. It has thought there are about 3,500 species of woodlice in the world. Of the 40 species found outdoors in Britain, Common Pygmy Woodlouse *Trichoniscus pusillus*, Smooth Woodlouse *Oniscus asellus* and Rough Woodlouse *Porcellio scaber*, are common just about everywhere.

Woodlice are sometimes given local names such as pill bugs and slaters. The pill woodlouse gets its name because it can roll itself up into a ball.

Woodlice like damp, dark places; hence, the floorboard, and can be found hiding in walls, under stones and in compost heaps, all throughout the year. Some species such as the common sea slater are only found on the coast. It easily becomes desiccated, so will hide in damp places during the day, especially in hot, dry weather.

A Woodlouse has 14 legs and an outer shell called an exoskeleton, averaging 10 mm in length. When a Woodlouse grows too big for its exoskeleton it has to moult to allow

Fig 99. Frass at the edges.

Fig 100. Woodlice scavenging.

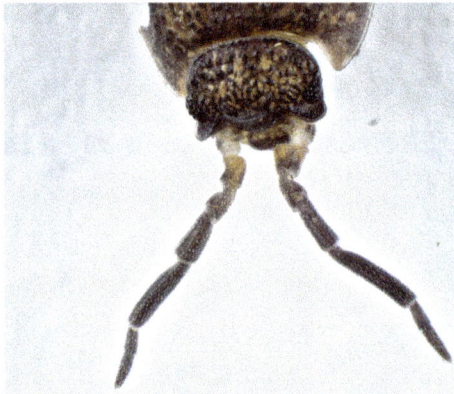

Fig 101. Head of Woodlice.
X 25 magnification.

Fig 102. Tail end showing
uropods. X 25 magnification.

a new shell to take its place. Moulting takes place in two stages, first, the back half is shed and a day or so later the front half falls off. They have a pair of *antennae* to help them find their way around, and two small 'tubes', called *uropods*, sticking out the back of their bodies. The uropods help them navigate and some species use them to produce chemicals to discourage predators. Most woodlice are found on land, but their ancestors used to live in water and woodlice still breathe using gills.

Woodlice eat rotting plants, *fungi* and their own *faeces*, but they do not urinate. They get rid of their waste by producing strong-smelling chemical called *ammonia*, which passes out through their shells as a gas.

After mating, females carry their fertilised eggs in a small brood pouch under their bodies. The young hatch inside the pouch and stay there until they are big enough to survive on their own.

A common Woodlouse can live for three-four years. Apart from man, its main predators are Centipedes, Toads, Shrews and Spiders.

The first Woodlice were marine *isopods*, which are presumed to have colonised land in the *Carboniferous* period of earth's history, over 300 million years ago.

Uninvited guests

Over the last bee-keeping season, I have collected various insects from honey samples that I have been given to try to identify the pollen source and some that have lived in the hive. I think that they are all correctly identified. If we have an insect expert amongst us please step forward!

Fig 103. This was found in a honey sample. With eight legs, this makes it a mite, possibly brought in from a flower visited by a Bumblebee, as it seems to be a Parasitellus mite of some sort, which are a pest of the bumblebee and can be seen on their body if heavily infested. They do not seem to cause honeybees a problem and are only found in the hive on occasions.

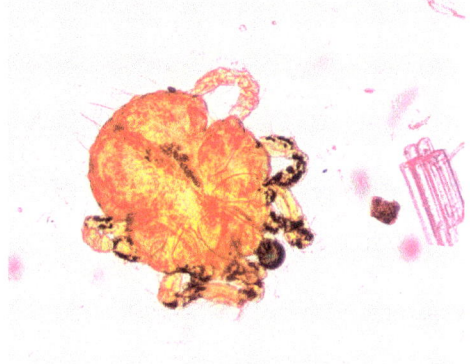

Fig 104. This was found in a honey sample. It is a bee louse, more commonly known as Braula coeca; once a pest, but now rarely seen due to Varroa treatment killing it off. Eggs are laid inside the honey cells before capping; the resulting larvae burrow out and eventually pupate. The adult emerges and likes to find the queen being a permanent member of the colony. They feed on the queen's saliva and are not harmful unless they become excessive, decreasing the efficiency of the queen.

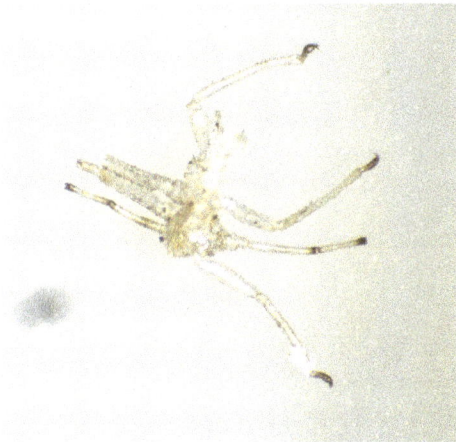

Fig 105. This was found in a honey sample. Due to its poor shape, I am unable to identify it, possible an opportunist! It looks like the cast cuticle of an insect. Does anyone recognize it?

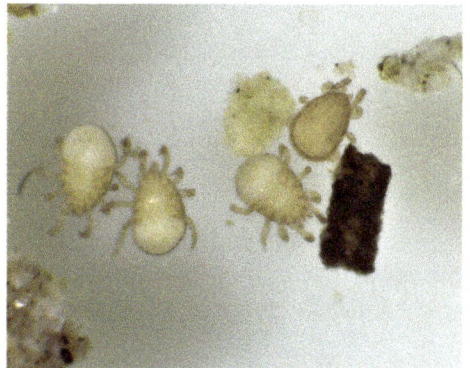

Fig 106. These are Pollen mites, taken from the bottom board; not normally a problem for honeybees. They will eat the pollen stores.

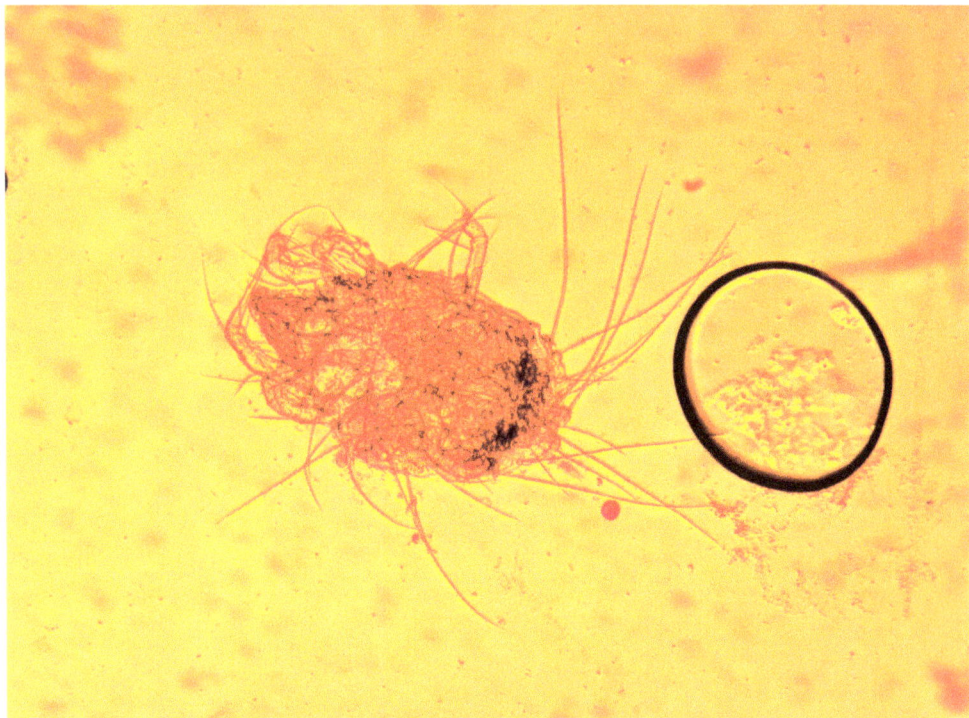

Fig 107. Acrine *mite taken from a honey sample, I have never seen this before. I think that the* Varroa *treatment is controlling these insects; the hive it was found in had been treated regularly for* Varroa. *Once the scourge of beekeepers it gains entry to the trachea via the first spiracle, where it bites & feeds from the bee's heamolymph. Blamed for the Isle of White disease in 1920, although this is now in doubt.*

Fig 108. *Taken from the bottom board, the dreaded* Varroa *mite, a member of the spider family. Public enemy number 1.*

Fig 109. *This is a* Wax moth larva *taken from the bottom board and will cause major problems and destroy a hive if infested in moderate numbers. There are two types found in the UK; the Lesser and Greater Wax moth.*

Over 90 differing mites infest honeybees in Europe, most do no harm.

Fig 110. Dead wax moth on bottom board.

Fig 111. Earwigs in the top of the super - they do not seem to venture into the hive and are ignored by the bees.

Fig 112. Lots of frass at the edge of the bottom board taken in late November - what is it? There is no evidence of something lurking above in the brood chamber.

Fig 113. Answer: A family of Woodlice has taken up residence.

During the day, woodlice hide in dark damp places so they are often found under logs, stones and flowerpots. At night, they move around in search of food, which is mainly rotting plant material. They do not seem to venture inside the hive and are ignored by the bees.

What the Microscopist DIDN'T want to see!

The annual visit from the bee inspector came in the first week of June this year. I have a very strong colony in a poly hive which over wintered well and has filled a super with honey. The second hive was a swarm collected in July last year which I re-queened, this also over wintered well but come early May I noticed a decline in it. A third batch was a swarm collected a week ago now in a poly nuke.

I believe that both of the hives were clean in May but at the last inspection, the swarm hive was a cause of concern.

By chance, the bee inspector was coming in 4 days time. Her sharp eye noticed the holes in the brood comb, these turned out to be emerging bees, but her nose was alerted by an unusual smell and on further looking in bright light the suspected evidence was revealed, *European Foul Brood* (EFB). Squishy brown centered grubs lying on the side of the cell. A secondary invader causes the smell, *Alvei bacteria*.

One of these was teased out and tested using a lateral flow device, and sadly, it proved positive.

Due to the small size of the colony and low infestation rate the decision was made to destroy the colony that night, so at 10 pm with all holes blocked and a board secured underneath I poured petrol into the crown board hole, refitted the lid and listened.

An immediate roar was heard followed by a slow decline until a single buzz was heard after about 30 seconds, one sound I do not wish to repeat again. The frames were burnt and hive scorched by the bee inspector and a 6-week non-movement of bees or equipment was then issued.

A standstill notice was issue and will remain until the bees have been looked at and checked out to be clear, the follow up is usually 6 weeks later, but can remain if the bees were not clear.

I eagerly wait for the next inspection to declare the other hives are clear. Prevention is paramount and the best way to help is to inspect your frames without the bees on it regularly and to wash your tools between hives, and the smoker afterwards with a strong solution of soda.

The causative agent of EFB is a *bacteria Melissococcus plutonius*, but another *bacterium*, *Paenibacillus alvae* is also associated with the grub's demise.

Samples are sent away to confirm the findings and the DNA is recorded to map the strains throughout the country.

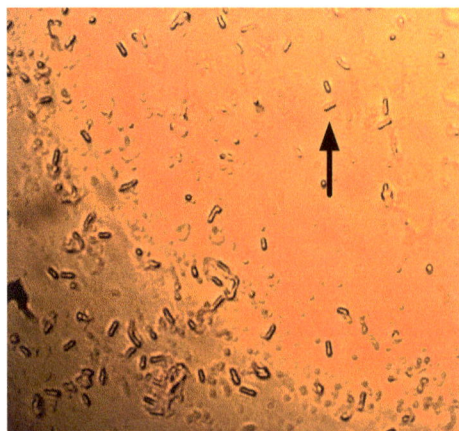

Fig 114. Spores of Paenibacillus alvae X1000 magnification, oil immersion.

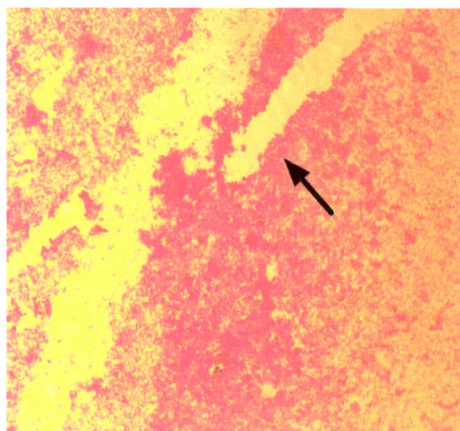

Fig 115. Melissococcus plutonius bacteria X 1000 magnification, oil immersion.

Imposter!

After inspecting my hives I was quietly de-robing outside of my garage when to my surprise a hornet buzzed me around the head, with my trousers down I was stuck for options, it was very quick and large, could it be an Asian hornet, it most defiantly did not look like our native one.

I trapped a few last year and froze them to show folks in the bees under the microscope days, so I feel quite acquainted with them.

I am sorry to say that it was pinned against the wall rather to heavily in my haste to capture it. On further inspection it did partially look like an Asian hornet with yellow legs and a black back, however, there was a large stinger with some floating tails underneath, panic over, it was an imposter!

This was the first time I had seen a Giant Wood wasp, (*Urocerus gigus Siricidae*) also know as Horntail a member of the Sawflies family, which can often be confused with the Asian hornet when flying at speed.

These insects are larger than the hornets being up to 45mm long, Hornets are 30mm long, and totally harmless the yellow legs and black banding give reason for confusion; however it is the ovipositor that is very long and the larger wings that really identify it.

They normal live in pine-forested areas.

Fig 116. Ventral view showing long yellow antennae and yellow legs.

Fig 117. Dorsal view.

Viruses and the honeybee

There are 24 known viruses that attack the honeybee, not all of these are found in Europe. The following article explains what a virus is, how they infect the bee and the common types of viruses in the hive. The word 'Virus' derives from Latin, meaning venom.

Viruses are neither plants, animals, nor bacteria, but they are the parasites of the living kingdoms. Although they may seem like living organisms because of their prodigious reproductive abilities, viruses are not living organisms in the strict sense of the word. A virus is one of the smallest units and can vary between 20 and 400 nanometres in size; a human cell membrane thickness is 7 nanometres thick and about 2500 nanometres in diameter (250 micrometers)

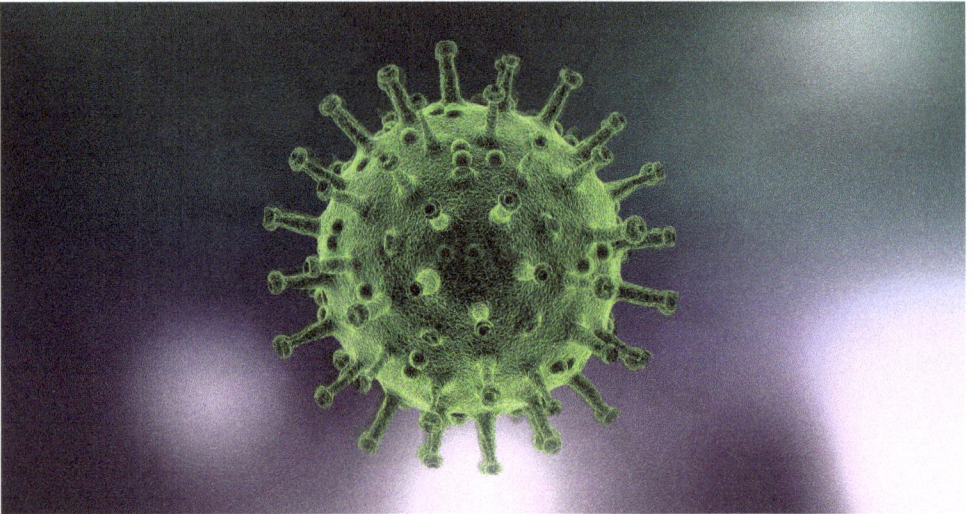

Fig 118. An electron beam magnification of an animal virus. Image by PIRO4D from Pixabay.

Without a host cell, viruses cannot carry out their life-sustaining functions or repro-duce. They cannot synthesize proteins, because they lack ribosomes and must use the ribosomes of their host cells to translate viral messenger RNA into viral proteins. Viruses cannot generate or store their own energy, but have to derive their energy, and all other metabolic functions, from the host cell. They also parasitise the cell for basic building materials, such as amino acids, nucleotides, and *lipids* (fats). Although viruses have been speculated as being a form of protolife, their inability to survive without liv-ing organisms makes it highly unlikely that they preceded cellular life during the Earth's early evolution. Some scientists speculate that viruses started as rogue segments of genetic code that adapted to a parasitic existence. There is evidence of viral codes in human cells and very large viruses have been found underground in caves that have been sealed for millennium.

All viruses contain nucleic acid, either DNA or RNA (but not both), and a protein coat, which encases the nucleic acid. Some viruses are also enclosed by an envelope of fat and protein molecules. In its infective form, outside the cell, a virus particle is called a virion. Each virion contains at least one unique protein synthesised by specific *genes* in its nucleic acid. Viroids (meaning "virus like") are disease-causing organisms that contain only nucleic acid and have no structural proteins. Other virus like particles called prions are composed primarily of a protein tightly integrated with a small nucleic acid molecule.

Viruses are generally classified by the organisms they infect, animals, plants, or bacteria. Since viruses cannot penetrate plant cell walls, virtually all plant viruses are transmitted by insects or other organisms that feed on plants. Certain bacterial viruses, such as the T4 bacteriophage, have evolved an elaborate process of infection. The virus has a "tail" which it attaches to the bacterium surface by means of proteinaceous "pins." The tail contracts and the tail plug penetrate the cell wall and underlying membrane, injecting the viral nucleic acids into the cell. Viruses are further classified into families and genera based on three structural considerations: 1) the type and size of their nucleic acid, 2) the size and shape of the capsid, and 3) whether they have a *lipid* envelope surrounding the nucleocapsid (the capsid enclosed nucleic acid).

There are predominantly two kinds of shapes found amongst viruses: rods, or filaments, and spheres. The rod shape is due to the linear array of the nucleic acid and the protein subunits making up the capsid, these attack bacteria; the sphere shape is actually a 20-sided polygon (icosahedron) and attack mainly animal life.

Diagrammatic view of how a virus infects a cell.

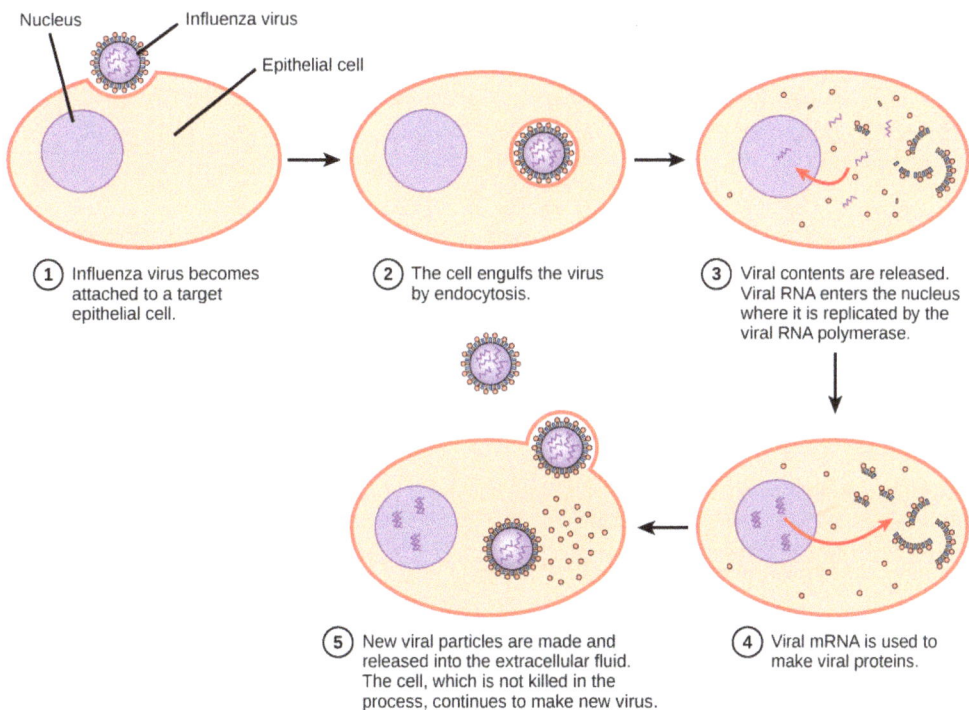

1. Influenza virus becomes attached to a target epithelial cell.

2. The cell engulfs the virus by endocytosis.

3. Viral contents are released. Viral RNA enters the nucleus where it is replicated by the viral RNA polymerase.

4. Viral mRNA is used to make viral proteins.

5. New viral particles are made and released into the extracellular fluid. The cell, which is not killed in the process, continues to make new virus.

Fig 119. Image provided by: OpenStax CNX. http://cnx.org/contents/e42bd376-624b-4c0f-972f-e0c57998e765@4.2. License: Creative Commons.org.

The nature of viruses was not understood until the twentieth century, but their effects had been observed for centuries. British physician Edward Jenner even discovered the principle of inoculation in the late eighteenth century, after he observed that people who contracted the mild cowpox disease were generally immune to the deadlier smallpox disease. By the late nineteenth century, scientists knew that some agent was causing a disease of tobacco plants, but would not grow on an artificial medium (like bacteria) and was too small to be seen through a light microscope. Advances in live cell culture and microscopy in the twentieth century eventually allowed scientists to identify viruses. Advances in genetics dramatically improved the identification process.

One of the main hosts of viruses in honey bees are the *Varroa* mite, they attack honey bee colonies as an external parasite of adult and developing bees, by feeding on hemolymph (fluid of the circulatory system similar to blood), spreading disease, and reducing their lifespan. Evidence suggests that *Varroa* and their vectored viruses affect the immune response of honeybees, making them more susceptible to disease agents.

Varroa have two life stages, phoretic and reproductive. The phoretic stage is when a mature *Varroa* mite is attached to an adult bee and survives on the bee's hemolymph. During this stage, the mite may change hosts often-transmitting viruses by picking up the virus on one individual and injecting it to another during feeding.

Viruses found in honeybees have been known to scientists for 50 years and were generally considered harmless until the 1980's when *Varroa* became a widespread problem. Since then, twenty-four honeybee viruses have been discovered and the majority of them have an association with *Varroa* mites, which act as a physical and or biological vector. Therefore controlling *Varroa* populations in a hive will often control the associated viruses and finding symptoms of the viral diseases is indicative of a *Varroa* epidemic in the colony

Viruses are microscopic organisms that consist of genetic material (RNA or DNA) contained in a protein coat. Viruses do not acquire their own nutrients or live independently, and can only multiply within living cells of a host. An individual virus unit is called a virus particle or virion and the abundance of these particles in a host is called the virus titer. A virus particle injects itself in to a host cell and uses the cells' organelles to make copies of itself. This process will continue without obvious change to the cell, until the host cell becomes damaged or dies, releasing large amounts of infective virus particles. All forms of life are attacked by viruses and most are host specific.

Honeybee Viruses

Viruses of the honeybee typically infect the larval or pupa stage, but the symptoms are often most obvious in adult bees. Many of these viruses are consumed in pollen or the jelly produced by nurse bees that are fed to developing bees.

Identification of a virus is difficult due to the small size of particles. Expensive and often uncommon laboratory equipment is required for accurate diagnosis. However, symptoms of some viral diseases are more visible, especially with overt infection. A lack of symptoms does not rule out the presence of a virus. Viruses can remain in a latent form within the host, acting as a reservoir of infection, complicating diagnosis and control, and only becoming an outbreak when conditions are right.

Sacbrood was the first honeybee virus to be discovered in the early 20th century and now has a recognized widespread distribution. It is perhaps the most common honeybee virus. This disease has been found in adult, queen, egg, and larval bees, in all forms of food, and in *Varroa* mites, suggesting a wide range of transmission routes. It often goes unnoticed since it usually infects only a small portion of brood, and adult bees will usually detect and remove infected *larvae*.

The disease causes *larvae* to fail to shed their final skin prior to pupation, after the larva has spun its cocoon. Infected *larvae* remain on their back with their head towards the cell capping. Fluid accumulates in the body and the colour will change from pearly white to pale yellow, with the head-changing colour first. Then, after the larva dies, it becomes dark brown with the head black

Deformed wing virus (DWV) is common, widely distributed, and closely associated with *Varroa* mites. Both the virus titers and prevalence of the virus in colonies are directly linked to *Varroa* infestations. In heavily *Varroa* infested colonies, nearly 100 percent of adult workers may be infected with DWV and have high virus titers even without showing symptoms.

Acute infections of DWV are typically linked to high *Varroa* infestation levels. Symptoms noted in acute infections include early death of pupae, deformed wings, shortened abdomen, and cuticle discoloration in adult bees, which die within 3 days causing the colony to eventually collapse. Not all mite-infested

Fig 120. Sac brood infected pupa, showing curled position. Open Government License. Beebase.

Fig 121. Adult bees with deformed wings resulting from DWV. Open Government License. Beebase.

Fig 122. Dysentery on the front of a hive is a symptom but not indicative of Nosema disease. Open Government License. Beebase.

pupae develop these symptoms, but all adult honeybees with symptoms develop from parasitized pupae.

Black queen cell virus (BQCV) is a widespread and common virus that persists as asymptomatic infections of worker bees and brood. Unlike other viruses that are associated with *Varroa*, BQCV is strongly associated with *Nosema apis* and little evidence supports its co-occurrence with *Varroa*.

Chronic bee paralysis virus (CBPV) was one of the first honeybee viruses to be isolated. It is unique among honeybee viruses in that it has a distinct particle size and genome composition. It is also the only common honeybee virus to have both visual behavior and physiological modifications resulting from infection. Symptoms of the disease are observed in adult bees displaying one of two sets of symptoms called syndromes. Type 1 symptoms include trembling motion of the wings and bodies of adult bees, who are unable to fly, and crawl along the ground or up plant stems, often clustering together. The bees may also have a bloated abdomen, causing dysentery and will die within a few days after displaying symptoms.

Fig 123. Bees with CBPV type two symptoms: greasy and hairless. Open Government License. Beebase.

Type 2 symptoms are greasy, hairless, black adult bees that can fly, but within a few days, become flightless, trembling, and soon die (Image 5). Both of these syndromes can occur within the same colony.

Transmission of the virus primarily occurs through direct body contact, although oral transmission also occurs but is much less virulent. Direct body transmission happens when bees are either crowded or confined within the hive for a long period of time (due to poor weather or during long-distance transportation) or when too many colonies are foraging within a limited area, such as a monoculture of sunflower with high honeybee colony density. In both instances, small cuts from broken hairs on an adult bee's cuticle and direct contact with infected adult bees spreads the virus through their exposed pores; if this occurs rapidly and enough adult bees are infected, an outbreak with colony mortality will occur. Faeces from infected bees within a colony can also spread the disease, and other transmission routes are still being investigated, including possible *Varroa* transmission.

Acute bee paralysis virus (ABPV), Kashmir bee virus (KBV), and Israel acute paralysis virus (IAPV) is a complex of associated viruses with similar transmission routes and affect similar life stages. These viruses are widespread at low titers and can quickly develop high titers due to extremely virulent pathology. Frequently associated with colony loss, this virus complex is especially deadly when colonies are heavily infested with *Varroa* mites. These viruses have not been shown to cause symptoms in larval life stages, but show quick mortality in pupae and adult bees.

Summary

Most pathogens invade the digestive system through oral ingestion of inoculated food. These pathogens infect the mid gut *epithelial* cells, which are constantly being replaced and are protected by membranes and filters which confine the pathogen to gut tissues. Parasites that infect gut tissue like *Nosema apis* and *Nosema ceranae* can create lesions in the epithelium that allow a virus like BQCV to pass into the hemolymph and infect other cells in the body. In contrast, the external parasite *Varroa destructor* feeds directly on bee hemolymph providing an opening in the cuticle for viruses to enter. *Varroa* mites can directly transmit many viruses, such as DWV, those in the acute bee paralysis virus complex, and slow bee paralysis virus. Other viruses, like Sacbrood, have been detected in *Varroa* mites but *Varroa* has not been shown to directly transmit the virus. Some viruses, like DWV, have been shown to directly multiply in *Varroa* mites, however in most cases we do not know the exact relationship of *Varroa* mites to viruses or enough about how transmission occurs from mites to bees. Knowledge about the presence, role, and transmitting routes of these viruses in native bees, and other potential non-*Varroa* transmission routes is also lacking in detail, complicating recommendations for control.

Frischy tales!

Karl von Frisch was awarded the Nobel Prize for his research into honeybees; however, he did not start out life as an entomologist. His first aim was to become a doctor but his uncle who taught him anatomy at university, suggested that he changed direction to become a zoologist, the rest is history.

However, did you know he dedicated his doctoral thesis to the study of light perception and colour changes in fish? Frisch discovered that minnows had an area on the forehead that is filled with sensory nerve cells, which he called a "third, very primitive eye." He showed that Blind Minnows could react to light by changing colour in the same way as minnows with sight. Frisch's discovery contradicted the common belief of his time that fish and all *invertebrates* were colour blind, and with this he stirred serious discussion among scientists, which held that the survival of most of the animal species depended on the development of their senses. He argued that animals adapt their behavior to better suit environmental conditions.

Frisch also studied auditory perception of fish. Again, contrary to the established belief of the times that fish could not hear, Frisch argued that they could, and designed a series of experiments to prove his point. He conditioned fish by pairing the sound of a whistle with the delivery of their food. He discovered that fish responded to the sound even when the food was absent. He showed that fish could hear, and later proved that the auditory acuity and sound-distinguishing ability of fish is more developed and superior to that of humans. This work helped to prepare him for his most famous discoveries. A recently published book that I would recommend called 'What a Fish Knows' by J Balcombe sets out to educate the public on just how advanced the species are. I am afraid you will be thinking twice about eating your next fishy meal after reading it; they know a lot more than we can understand and have evolved to adapt to the own environment, not ours.

Frisch however became famous for his study of honeybees. He first decided to prove that bees could distinguish colours, and started from his assumption of the adaptive function of the behavior. He argued that the bright colours of flowers developed to attract bees for the purpose of pollination. The logical conclusion would be that the bees could perceive those colours. To prove his point Frisch conditioned bees to respond to the blue-coloured objects, which contained sweet water. After removing the sweet water, bees would still come to the blue-coloured objects, proving that they could distinguish colours.

In 1919, Frisch demonstrated that bees could be trained to distinguish between different tastes and odours. He discovered that their sense of smell is similar to that of humans, but that their sense of taste is not so sophisticated.

Frisch's most distinguished discovery was that of the "waggle dance" of the scout bee. Through his previous experiments, he had noticed that scout bees somehow "communicate" the exact location of food to the other bees. For the purpose of study, Frisch constructed a glass honeycomb so that the bees could be observed from all sides. He noticed that scout bees, after returning from a successful food search, conveyed their finding to the other bees by performing two types of rhythmic movements—circling and wagging. The round circling movement was used to indicate relatively close sources of food, while the more complex form indicates food sources at greater distances. The latter became known as the bees' "waggle dance."

In 1949, Frisch proved that bees could perceive polarized light, which helps them to navigate through space. They use the sun as their main compass, and in the absence of the sun, they memorise patterns of polarization of the sky at different times of the day. They could also memorise the location of certain landmarks in nature.

There has been a recently published biography of Frisch 'The Dancing Bees: Karl Von Frisch and the Discovery of the Honeybee Language' by Tania Munz'. Frisch was 1/8th Jewish, however because the Nazis considered that bees were essential to the pollination of crops, von Frisch s research was deemed critical to maintaining the food supply of a nation at war. The bees, as von Frisch put it years later, saved his life.

Let us hear it for the Wax Moth

The wax moth gets bad press as far as the beekeeper is concerned. It lays it *lava* in the wax comb from which it then proceeds to devour the wax, leaving a vast number of tunnels as evidence; this can lead to the death of the colony if the investigation is a heavy one.

The Greater Wax moth, *Galleria mellonella*, is the more destructive and common pest whilst the Lesser Wax moth, *Achroia grisella*, is both less prevalent and less destructive. As usual in nature there always seems to be a balance between host and pest, otherwise the host would soon die out and take the pest with it.

Not all relationships are symbiotic or even instantly recognisable nor do they appear good for the host, but the Wax moths do clear up old colony sites left by swarming bees in the wild, removing all that remains including not only the wax, but all the silk and waste that the honeybee *larva* has deposited over time in the comb; think of them as super cleaners!

One more impressive use that scientists have put these critters to is in aiding medical research for the greater good of human kind. Wax moth lava change colour when they

Fig 124. Newly emerged Wax moth. By kind permission of Richard Ball.

are sick; this indicator makes them suitable when testing for bacterial concentrations that can be related to humans; sometimes it is 1,000 *bacteria* per millitre that causes a problem or 10 million *bacteria*. From this data, the infection rate then can be defined.

As a practical example, using the method mentioned above we are recommended to wash fruit off before we consume them; this does not remove all the bacteria but does dilute them down to a more manageable level for our bodies. Could their *larvae* be the next super food? We in the west do not seem to eat insects yet, but with the world exponential growth of people we may well have to!

Fig 125. The resulting damage done by the wax moth silk, black specks are frass. By kind permission of Richard Ball.

Earwigging

I have often seen the odd Spider, Woodlice and Fly snooping around the hive throughout the season but families of earwigs have taken up residency in the super frame runners and the crown board; none seems to venture below and the bees seem not to mind their presence. What do we know about this much-maligned critter? The Earwig is thought to get its name from people fearing that earwigs crawled into your ear to lay their eggs. Although this is not the sole intention of the Earwig, it is certainly thought to be possible as they like narrow, warm spaces such as the ear canal. However, the days of humans sleeping on the scattered straw bedding are now long gone!

People often see this insect and think that it is a one of the strangest looking bugs. These bugs have what appear to be pincers extending from their *abdomens*; these pincers are not used to aggressively attack people; however, if disturbed, they can

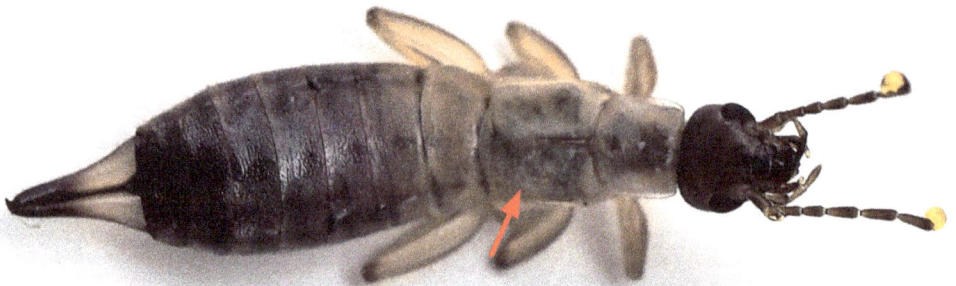

Fig 126. The wings are beneath casings.

latch onto skin leading to a slightly painful pinch but this is not common. Male pincers are generally larger than the female ones; it is believed that they use these appendages for holding food.

There are four species of Earwigs in the United Kingdom, depending on the species, adults range in size from 5-25 mm. They are slender insects with two pair of wings and some species produce a foul smelling liquid that they use for defence. Outdoors, Earwigs spend the winter in small burrows in the ground and in spring the female lays up to 80 eggs in the burrows where she tends them until they hatch within a couple of weeks. Female earwigs are known to be extremely protective of their young, often watching over them until they have reached their second moult (Earwigs moult five times over the course of their lifetime). Earwigs also produce a *pheromone*, which scientists believe is the reason why they cluster together in large numbers. Immature Earwigs (*nymphs*) resemble the adults except they do not have wings.

Earwigs are attracted to lights so they can become a nuisance on porches and patios on summer evenings. Come the morning they tend to be found hiding under things; they tend to move into homes to find food or because of a change in weather; they are relatively fast moving running away quickly when the ground material is moved.

Fig 127. Hiding between super spacing's.

Earwigs are nocturnal animals that often hide in small, moist crevices during the day, and are active at night hunting, as they are omnivorous animals meaning that they will eat almost anything they can find. Due to their small size, Earwigs have a number of natural predators wherever they live in the world. Amphibians such as frogs, newts and toads are among the most common predators of the Earwig along with birds and other larger insects such as beetles. Earwigs are insects of the order *Dermaptera* stemming from the Latin word, *derma* meaning "skin" and *ptera* "wings" thus "skin winged" from the appearance of the front wings. To remove Earwigs, provide shelter in the form of inverted pots. Pack these loosely with dried grass and place the pots on the top of canes situated among the plants. The earwigs can be disposed of once caught, making the use of insecticides unnecessary. They have however earned a place in the English language: "To earwig" is a slang verb meaning either "to attempt to influence by persistent confidential argument or talk" or "to eavesdrop".

Fermentation blues

When I was an apprentice, my foreman, Alfie, 'the old boy', used to say that you could not buy experience at Woolworths! Well now that I am, the 'old boy' and Woolworths went several years ago from Barnstaple I must admit that I am still trying to gain that experience. I have a few jars of last year's crystallised honey in the cupboard and I noticed some sticky mess on the shelf. On investigating, I found one jar that had big bubbles in; the other two had only a few small minor bubbles on the top surface. On opening, the offending jar there was no noticeable smell and on tasting no off notes; it went in the bin. The other two I scooped off the top and re liquefied them; they tasted OK so I have now turned them into mead instead. I first boiled the honey and water to kill any remaining wild yeast cells, the proof will be in the tasting in two or more

Fig 128. The offending jar! Note the large bubbles at the top.

years time. Update: I got a second prize in our local competion!

Why does honey ferment?

The yeasts responsible for fermentation are endemic throughout our environment. The risk of fermentation is dependent on both the moisture content and yeast spore concentration within the honey and on the temperature at which the honey is stored. Honey with less than 17 percent water will not ferment in a year, irrespective of the yeast count. Between 17.1 and 18 percent moisture, honey with 1000 yeast spores or less per gram will be safe for a year. Above 19 percent water, honey can be expected to ferment even with only one spore per gram of honey, a level so low as to be very rare.

Since fermentation is dependent on temperature, honey will not ferment when stored below 10°C or above 27°C, the only problem when storing it in a fridge at 4°C is that it becomes set.

The relative moisture content in stored honey can increase in one of two ways. Moisture is added to the honey by being absorbed from the environment when the lid is left off for some time or when some sugar is removed from the honey sugar solution through crystallization; this causes the relative moisture content of the remaining sugar solution to increase. In order to crystallise there must be something like pollen for the sugar to seed to; when this happens the fluid between the crystals becomes diluted by removal of the solids and there is an increase of up to 6% water content, this then allows for fermentation if the temperature is too high. A number of physical changes also occur within the raw honey changing its physical characteristics as it ferments. The bubbles are caused by *carbon dioxide*, one of the by-products of fermentation, by *yeasts* and *ethyl alcohol* as it grows and feeds on sugar. The *ethyl alcohol* may then break down into *acetic acid* (vinegar) and water in the presence of oxygen. The combined flavours of *yeast*, *alcohol* and *acetic acid* make the honey unpalatable. You can greatly retard the crystallization process by placing honey in the refrigerator or freezer. This makes the honey so thick that it is hard for the *dextrose* molecules to move to a crystal and leave the solution.

Know the signs of fermentation

When you see any signs of fermentation the top layer of honey should be scooped off and discarded, and then the remaining honey should be re-liquefied as soon as possible; any bubbles seen when reheating are evidence of carbon dioxide in the honey. The moisture content of the sugar solution will again be below 18.5% stopping the fermentation. When you have had to re-liquefy a jar of honey to stop it from fermenting, then you know that particular jar of honey could ferment again when crystallized and it should be used as soon as possible.

How about some honeydew?

Honeydew is a sugar-rich sticky waste product that is secreted by *aphids* like the greenfly when they feed on plant sap. When their mouthpart penetrates the *phloem*, the supporting tissue of plants, the sap is at high-pressure and is forced into their gut, causing the previously digested sap to be ejected out of the gut's terminal opening. This residue is called honeydew and is deposited onto the plant surface, which then can attract *fungal* spores from the air; these can cause sooty *mould*. The bane of all gardeners and possibly the reason for the historic potato famine in Ireland.

Honeybees will collect the honeydew from the leave surfaces, ants have been observed to collect it from the rear end of the aphid directly.

Technically, to be called honeydew it must pass through the bugs gut first as sap is just sap. Plant sap contains a large quantity of water, and in order to extract sufficient nutrients to survive, a large quantity of sap must be ingested. The *alimentary* tract of the *aphid* has a modification, referred to as the filter chamber, this allows nutrients to be concentrated in the midgut and small *intestine*; any excess water then bypasses the midgut and small intestine and is exuded from the *rectum* as honeydew. It attracts many kinds of insects, honeybees included, to feed on its sweet nutrients.

Fig 129. A greenfly on my red current leaf June 2015.

Aphids are dependent on sap for all their nutritional needs, including *protein*. However, plant sap is mostly water and sugar with just a fraction of *protein*—about 1 to 2 percent of the volume. The insect must eat large amounts of sap to get enough *protein*, so what comes out is very similar to what went in—minus a few *amino acids*, however they cannot live on sap alone!

There is a unique symbiotic partnership inside of the cells its self in the gut of *aphids*; they have teamed up with the bacterium *Buchnera aphidicola* in return for a safe haven and food; they help by producing essential *amino acids* which are lacking in the plant sap. Without these *bacteria* the *aphids* will die. The honeydew that is collected by honeybees is processed into a dark, strong honey (honeydew honey) which is highly prized in parts of Europe and Asia for its reputed medicinal value; it attracts a premium price in excess of £15.00 per jar. Like normal honey, the taste and colour will vary according to the plant source. The additional amount of honeydew in your honey depends on the plant species that live nearby, the climate, and the local weather. If floral nectar is plentiful all year long, honeydew collection will remain insignificant.

Unlike pure honey, which can ferment if the water content is too high, honeydew honey seems only to displays mould when it is left to long.

Natural trapping!

There are some concerns about trapping insects using liquid baits in the trap that do not allow for separation of the victims, allowing them to drown. The alternative slant can be seen in the pictures, albeit not my design or chosen method! My four hives sit behind a fence and backing onto two pear trees. I dutifully reduced the crop after the June drop, calculating for a few disfigured and course skinned ones, plus allowing for the odd mouldy and bird/insect eaten ones. The branches on the Conference tree were heavily populated this year, so a good crop looked promising.

The Wasp population has been heavy in Devon this autumn. I reduced my hive entrances down and witnessed a few dead and bee/wasp fights in the tray I keep beneath the entrance; the bees were copping admirably.

However, little did I realise that the reason for not bothering the bees this year was the sweet tasting fruit of the nearby pears. I did put a trap up to monitor for the Asian hornet, none found, but plenty of the native species, they all much preferred the fruit.

I reckon that about over fifty percent of the fruit has been damaged in some way. The pear ripens from the inside out, unlike the apple, so you have to pick them at the right time. The first tree yields a buttery type of pear, that tastes like honey (name unknown) and these pears keep for one week only.

The Conference is a keeping pear and is not picked until later, that is if there are any left. In practice, you wait until a few pears drop off the tree of their own accord and then harvest the remaining pears over a couple of weeks. They are ready for picking when they can be easily lifted off the tree. Not much chance of that happening. The wasps seemed quite oblivious to humans; drunk on fermenting fruit perhaps, if that is

Fig 130. Stage one, the making of the entrance; Fig 131. Stage two, the shelter; Fig 132. Stage three, the shell.

possible. They did seem to sit on leaves and groom themselves after being inside the pear. The question is, do I pick them now and do I wear a wasp suit? I once cut into a wasp nest with a hedge trimmer earlier in the year; I did not know it was there; I got fifteen stings on the back of my head, ouch!

Fig 133. One of the culprits.

Fig 134. An attempted forced entry by a wasp.

Drone congregation areas

It is widely evidenced that certain parameters must be obtained for successful mating to take place, but on the other hand, some bee scientists differ in their opinions as to whether drone congregation sites actually, or at least uniformly, exist.

The country parson, Gilbert White noted in his diary the phenomenon, we now refer to as a drone congregational area, in Hampshire, England in 1792, a site still reputed to be active.

Virgin queens and young drones fly some distance to reach these specific sites called drone congregation areas, where they gather, some 10–20 metres aloft, to pair in flight.

Drones start flying to these aerial rendezvous sites at around noon onwards, which is about one hour before the queens begin to arrive. Usually there is a horde of drones circling at each site by the time the first queen arrives.

When a young queen bee sallies forth to get inseminated, she travels normally without a retinue of worker bees for her protection. Because a virgin queen mainly flies alone, it is the riskiest time of her life as she is easy prey for predators. It is a moot point whether she finds the DCA using an inherited ability to recognise where drones are likely to establish a concentration, or whether she can detect their distinct flyways and follows them, or simply uses their drone *pheromones* to find the congregation by scent. Further research will give us the answer.

A virgin queen usually conducts just one mating flight and she keeps it brief, mating with 10–20 drones. The reason for multiple mating is to offer genetic mix to the hive and to limit inbreeding with a brother. It is known that some queens will make a second mating flight if she has not mated with enough drones.

Some sites where the virgin queens and sex-ready drones meet to mate appear to be stable from year to year. Most of what we know about drone congregation areas comes from the work of two brothers, Professors Friedrich and Hans Ruttner, who worked in Austria in the 1960s and 1970s, and their successors, Professors Gudrun and Nikolaus Koeniger, who have continued the investigations (in both Austria and Germany) to the present day.

These researchers have discovered that the "hook-up" sites of queens and drones can have remarkably distinct boundaries. In one location, for example, the Ruttners found that when they displaced an airborne queen, confined in a cage held aloft by a helium balloon, by only 30 meters within a drone congregation area, it often shrank by tenfold the number of drones hovering around the caged queen. They also found that in the mountainous regions where they conducted their studies, queens and drones appeared to orientate to their congregation areas by flying toward low points on the horizon line.

It may be that the drones may perceive as the directions of maximal light intensity. It may also be that drones continue orienting in flight in this manner until they reach a location where the intensity of light on the horizon is uniform. Where polarised light boundaries form they tend to circle.

How or whether drone congregation areas form where the countryside is flat remains a mystery. It may be that drones are distributed rather evenly over flatlands and that they congregate only when they detect the alluring scent of a queen and orient upwind to its source.

Two other inquiries about the mating habits of honeybees looked into the density of their mating sites and the distances that queens and drones will fly to reach them.

Fig 135. A mature drone, note the curved abdomen, large thorax and large eyes.

One intensive search conducted near Erlangen, in southern Germany, found five drone congregation areas within a circular area that covered about 3 square kilometres, with a density of approximately 1.6 congregation areas per square kilometre, with queens mating on average 2–3 kilometres from their homes and drones travelling 5–7 kilometres or more to find a sexually receptive queen. Koenigers claim the reverse – queens go further to avoid mating with related drones.

The queen is at risk for longer by flying further but drones have longer on-station if their back and forth flight time is minimised. This contradictory evidence is possibly due to different bees in different areas, just as DCA evidence varies from Carniolans in Austria to possibly *Mellifera* in north Germany and hybrids in England. Perhaps the most impressive evidence of drones making long-distance mating flights comes from a massive mark-and-recapture study conducted in the Austrian Alps by Friedrich and Hans Ruttner in the mid-1960s. They began by going to these apiaries and labelling thousands of drones, each with a colony specific paint mark. Next, they captured drones at two of the six known drone congregation areas in the region. Their capture method worked as follows: they lofted a queen in a small plastic cage suspended from a helium-filled balloon; they waited until a crowd of drones was circling around her, and then they slowly lowered her to where they could collect the queen-baited drones using a long-handled insect net.

Amazingly, at drone congregation area C, which sits in a high valley in the centre of their study site, they captured drones from 18 of the 19 apiaries in the region. The only apiary not represented among the drones captured at congregation area C was apiary 9, which was only 1.6 kilometres from this congregation area but was separated from it by the Seekopf mountain, rising more than 300 metres.

The Ruttners also reported how many of their captured drones came from each apiary, and from their data I have calculated the average distances flown by the drones they captured at congregation areas B and C: 3.0 kilometres and 2.3 kilometres, respectively.

The longest mating flight they detected was an excursion made by a drone from apiary 17 to congregation area C. He flew either a 3.9 kilometre beeline route over the mountains or more likely went around the mountains via an approximately 6 kilometre curved route down the long valley leading to the lake. These findings about the impressive mating flight distances of drones in the Austrian Alps are supported by Donald F. Peer the findings of in the 1950s working in Ontario, Canada. He studied the mating range of honeybees by introducing colonies to a region covered with vast coniferous forestlands that contained no colonies other than his experimental ones. He established an apiary stocked with 20 colonies that produced only drones carrying the *Cordovan allele*, a recessive colour mutation. He also set out small colonies and mating *nuclei*, each of which had no drones but contained a virgin queen that was genetically marked by being *homozygous* for the *Cordovan* mutation.

To get data on mating flight range, he placed his mating nuclei containing virgin queens at various distances from his full-size colonies containing drones. He found that none of the 22 queens that were separated from the drone-source colonies by 19.3 or 22.6 kilometres mated successfully but that most of the queens separated from the drone-

Figure 136. Drones being attracted to a lure in mid July that is baited with queen pheromone 9-oxo-2-decenioc acid.

source colonies by 16.0 kilometres or less did mate successfully, and only with males carrying the *Cordovan allele* (hence from his apiary).

Although Peer's impressive results reveal maximum mating ranges, not typical ones, the fact that honey bees have such huge mating ranges indicates that strong out breeding is almost certainly the rule for *Apis mellifera*.

A drone congregation area in Austria studied in detail was reported (ref) to have ~10,000 drones from 240 different colonies. If all colonies were equally represented, this would amount to ~42 drones per colony. If the average number of matings per queen is 13, then the probability of an individual drone mating with an individual queen who enters a drone congregation area is about 13 in 10,000 or 0.0013%: a very slim chance.

In terms of genetic representation it is in the interests of every colony for its drones to mate with as many virgin queens as possible, but if we assume an individual colony seeks to just maintain its current genetic representation in a closed population it can be calculated that each colony's drones must achieve an average of 6.5 matings with other colonies' queens per year – assuming queens are replaced bi-annually.

240 colonies/2= 120, therefore drone matings achieved in a given DCA will be 120x13 (on average), i.e. 1560 per annum. 1560/240 contributing colonies = 6.5.

The drone's flight strategy is more limited in range than that of workers, as they carry only relatively small amounts of honey on their mating flights as well as small glycogen reserves, and as they are not providing themselves with nectar in the field as workers do. It is also important for each drone to be in peak condition in an attempt to gain a competitive mating advantage. Successful colonies require a difficult combination of quantity and quality in its drone pool. If we also assume that in the absence of bee-keeper intervention or significant disease attrition every geographic area already hosts an optimal honeybee population spread across numerous colonies, then we can predict that on average, as many colonies will fail each year, as there are swarms.

As it is not at all obvious which colonies will succeed or fail at the time of queen mating and as the opportunity for any drone to mate is so slender, every drone will vigorously pursue every queen bee in need of mating that enters a drone congregation area.

It has been noted that drones of different races tend to fly at different heights, possible to stop cross mating.

As drone quality and fitness is of the utmost importance to the colony for the dispersal of its genetic legacy, considerable time and resources are invested in the development of the best possible specimens.

Odours are detected by the eighth *annuli* segments of the *antennae*. In addition to having extra *annuli*, drone *antennae* have 50% more surface area than worker *antennae* and over 6 times as many olfactory receptors.

Honey bee genetics: The drone's role

A brief description of honeybee genetics setting out how the bees reproduce and why the drone has no father; the drones also have no sons, but at most, they have grandsons!

Chromosomes are long structures that are found in most cells. They contain the *genes* of an organism (humans have about 24,000 *genes*, bees have about 15,000 *genes*). Most animals normally have two sets of *chromosomes*: one set comes from the mother and one from the father.

They are called *diploid* (di means two, and ploid stands for *chromosome*). Human beings have 46 *chromosomes*: we get 23 from our mother's egg and 23 from our father's *sperm*. Bees have a different number of *chromosomes*. Females, workers and queens, have 32; 16 are contributed by the queen's eggs and 16 come from the drone's *sperm*.

Since drones hatch from unfertilized eggs, they only have the 16 *chromosomes* that were in the egg. Drones are *haploid* (from the Greek word haplos meaning single).

Haploid males are a characteristic of the order, *Hymenoptera*, because they only have one set of *chromosomes*. The reproduction process in which the offspring develops from unfertilized eggs is called *parthenogenesis*, meaning asexual.

The egg can only carry half of the queen's 32 *chromosomes*, so she can only pass on half of her *genes* to her offspring; *genes* carry genetic material from one generation to the next. When the egg develops, it divides into 2 parts, splitting the 32 *chromosomes* in half.

This process is called *mitosis*. Each egg contains a unique collection of her *genes*, so each egg is different. Drones, on the other hand, only have 16 *chromosomes* to begin with, so each *sperm* must contain all the *genes* of the drone. This means that each *sperm* from a single drone is identical. They are clones.

This is different from most other animals, where each *sperm* is unique, just as each egg is unique. If we use humans as an

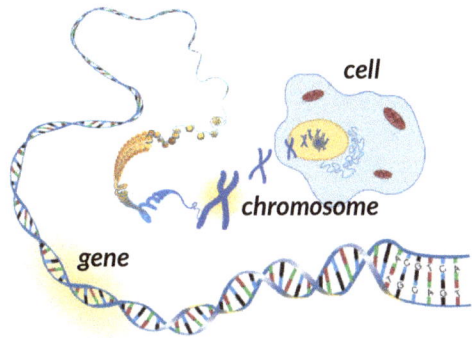

Figure 137: DNA, adapted from National Human Genome Institute. Reprinted by kind permission of Ryan Evans.

example, the females have two sex XX *chromosomes* both the same, and are called *Homozygous*, whereas males have two different sex *chromosomes* XY. This is called *Heterozygous*.

Depending on what sex *chromosome* the *sperm* cell has X or y, determines the sex of human babies.

In bees, the female carries two copies of the same *chromosomes* and are *homozygous*. However, the drone has only one set of *chromosomes* his genetic make-up is described as *hemizygous*.

Diploid drones can occur from a fertilised egg, according to Woyke (1963a). After the *diploid* drone *larvae* hatch from the eggs, they are not reared by the worker bees, but are eaten by the nurse bees within 6 hours after hatching. Professor Woyke in 1963 elaborated a method to rear *diploid* drone *larvae* to the *imago* stage.

Pair of *genes* for the same characteristic is called an *allelomorphic* pair, which is usually shortened to *alleles*.

Sex determination in the honeybee is controlled by about 15 sex *alleles*: coded for example: a, b, c, d, e, f, g, h. Normal drones developed from unfertilized eggs, with *allele* a, or b, or c, or d...develop into haploid drones.

Fertilized eggs usually develop bee females (workers and queens), in which the sex determining *alleles* are different (ab, ac, ad, etc).

Drones that have two sets of *chromosomes* but the same version of the sex *allele* are in effect haploid having only one sex *allele* contributing genetic material where there should be two different. From these *homozygous* eggs, aa, bb, cc ... *diploid* drones can develop.

After the *diploid* drone *larvae* hatch from the eggs, they are not reared by the worker bees, but are eaten by the nurse bees within 6 hours after hatching. Professor Woyke in 1963 elaborated a method to rear *diploid* drone *larvae* to the *imago* stage. See figure 54.

The other odd thing about bees is their habit of multiple mating. The queen is quite promiscuous and mates with from 10 to 20 drones, usually in one or two mating flights over the course of a couple days. The *sperm* is stored for years in an organ called the *spermatheca*.

Actually, the *sperm* from one drone is more than enough to fill the *spermatheca*. Therefore, it seems the queen goes out of her way and takes great risks to mate with so many drones, just to create extra genetic diversity for her colony.

It is thought that one of the reasons why the queen mates with so many drones is due to her trying to obtain the rare *alleles* that are found amongst the drone groups, these could help to combat new diseases in the future.

Figure 138: Diploid drone, Woyke 1977.

In addition, since sub-families of worker bees (share a common father) tend to specialise in performing certain tasks in the hive, a diversity of fathers may enable the colony to perform more efficiently. There is well-researched evidence that colonies headed by multiple-mated queens outperform single-mated queens with a dramatically better chance of winter survival.

The **phenotype** of an individual honeybee, a colony, or a population, is the set of observable characteristics such as size, colour, honey production, wintering ability and defensiveness. It is these characteristics that the bee breeder aims to alter. Longer term changes also occur but take time before a pattern of becomes measurable.

The **genotype** (the "genetics") of a bee or colony is the set of inherited genetic instructions encoded in its DNA. These can be either dominant or recessive. A recessive *allele* does not become a trait unless both copies of the gene, one from the queen and drone, are present. If one dominant *allele* and one recessive *allele* are present, the dominant *allele* trait will be expressed. When we refer to two bees having different "*genes*," what we really mean is that they have two different forms (variants) of the same gene. Both in natural selection and in traditional selective breeding, selection is applied to the expression of the *phenotype*, rather than the *genotype*, since it is the *phenotype* that directly interacts with the environment, and is an observable characteristic

Selection of the honeybee was not strongly influenced by humans because basic bee reproduction was not understood until about 1850 after Langstroth developed the movable frame hive. Suddenly beekeepers not only understood bee reproduction, they could also manipulate the hive and control the queen.

Within a colony, there are usually 7 to 10 *subfamilies* because of the queen mating with up to 20 different drones. Not all of the *sperm* from up to 20 matings is stored. Since all the *sperm* produced by a drone are genetically identical, each subfamily is composed of sisters that are more closely related than full sisters of other animals are. Often called *super sisters*, they will have three-quarters of their *genes* in common.

Despite the complicated family structure, the basic principles of genetics still apply to bees. On rare occasions, a *gene* entering an egg or *sperm* has changed somewhat and will have a different effect than the original *gene*.

The process of change is called, *mutation*.

Mutations and their effects

More than 30 specific visible mutations have been described in bees. Generally, these mutations produce a striking effect. Known mutations affect the colour of, shape, and presence of eyes, the colour and hairiness of bodies, the shape and size of wings, and nest-cleaning behaviour. Most mutations are recessive, and are first observed in drones.

The reason for this is that they do not have paired *chromosomes*, only single, so no dominant gene can be preferentially expressed.

Feral identity parade

In June, I was asked to look at a wild colony of bees that have resided low down,
in a wooden clad wall of a house. The
owner reported that feral bees had been
in residence on and off for over 25 years;
colonies will often stay in one location
for up to 5 years then swarm, allowing
a new swarm to take up residence once
the wax moth has done it cleaning role.
Were these truly wild feral bees, native
black bees or a hybrid mix? The property
is in a small hamlet surrounded by farming and natural heathlands. I collected a
few bees to examine under the micro-

Fig 139. A sample worker bee,
note the colour of the pollen.

scope and took pollen samples taken from their rear legs.

The first impression was that some of them were on the small size and the colour of the *abdomens* varies from yellow to black. Their nesting area was about 10 centimetres across and about a metre square; I had no means of seeing which way the comb went internally. The colour difference indicated the same queen but different fathers due to mixed mating that is normal for the *Apis* species.

Fig 140, 141. Same mother, different fathers. X 30 magnification.

On looking at the anatomy features to try to discover if they where native black bees, the wing measurements were taken to give a ratio of length between two veins and discoidal shift, an angular measurement usually a plus or minus angle, again using 2 different veins. The average ratio was 1.8 with a negative shift. Next, the hair length was measured; they had short hairs at 0.3-0.5 millimetres. The final check was the bandwidth across one of the abdomen segments; this was small, less than half of the surface area. On looking up the data, it would appear these are locally adapted honeybees, mainly of the Apis mellifera type that have interbred with others. As always, there were a few darker bees in the sample that conform more to the Native species. I am not aware of any native colonies locally. A quick health check was carried out, a sample was examined for Acrine mites, none were detected; most beekeepers now treat for Varroa mites, which also kill off any Acrine mites, but as these have not had any treatment, some might be expected to have been found. Good news, no Varroa mites present. However, most feral colonies that have been inspected have all had Varroa present. I really needed to look inside the wall. Next, their abdomens were ground up with water to make a slide in order to check for Nosema spores, again none was found. The pollen was mixed in alcohol, stained and a slide made. On examination the white colour on the workers' legs, give an approximate idea of the possible plants visited, plus the time of year and local flora.

The conclusion of flowering types was as follows, Maize, Iris, Valerian, Knapweed and Bindweed.

The overall verdict was that the colony was very healthy and surviving naturally without the aid of the beekeeper.

Good evidence for only using the locally adaptive bee species, whatever their parentage. Sadly, not all wild colonies are able to find new suitable nests, luckily for them the owner was very happy and proud to let nature be!

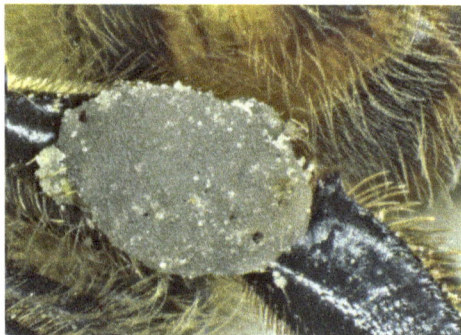

Fig 142. White-grey pollen pellet. X 40 magnification.

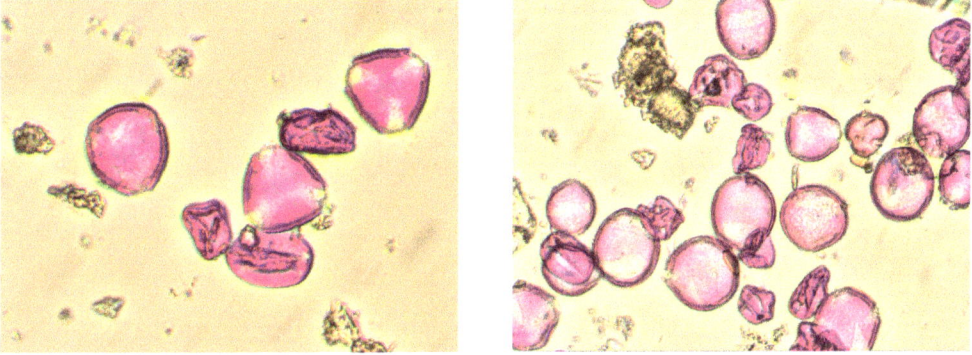

Fig 143, 144. Mixed pollen samples. X 400 magnification.

Morphometry or *Morphology?*

Morphometry is the precise study of anatomical characters by measurement and morphology is merely the study of form and structure; both words are used throughout the world, for measuring bee bits! Measurement of wing vein characteristics is used to establish race and breeding purity in honeybees, with the intention of using the data for selective breeding purposes. This can easily be done by the beekeeper using a variety of methods. I have used a dissecting microscope and a bespoke camera with suitable software. 30 dead bees from 2 hives were collected in early November, from outside their hive entrances. One hive was a new nuc raised late last year from Cornwall and the other a local swarm collected this year; both sets of bees look much the same in colour. The right forewing was removed from each bee.

The Racial types and strains of honeybees have distinctive body characteristics that can help to distinguish both type of bee and purity of breed. Of these two aspects, the one of greatest importance is purity of strain or more precisely the degree of hybridisation, the lower the better. These methods are all of secondary importance to 'colony assessment' characteristics and should be used to refine partly selected strains rather than as a direct descriptor of race. There is no point in propagating 'bad' or undesirable behavioural traits regardless of how 'pure' the strain is.

The *phenotype* of an individual honeybee, a colony, or a population, is the set of observable characteristics such as size, colour, honey production, wintering ability, and defensiveness. It is these characteristics that the bee breeder aims to alter.

The *genotype* (the "genetics) of a bee or colony is the set of inherited genetic instructions encoded in its DNA. These can be either dominant or recessive, depending on the mix from each parent.

Both in natural selection and in traditional selective breeding, selection is applied to the expression of the *phenotype*, rather than the *genotype*, since it is the *phenotype* that directly interacts with the environment.

Measurable characteristics

General appearance is a simple and obvious characteristic that most would agree on, but it is subjective in nature and you should be wary of 'seeing what you desire to see'.

Body colour is only important if describing a bee with a high degree of "purity of strain". Colour in itself is not a positive indicator, but can be used to rule out certain hybrids. I have experienced a wide variety of colour in my hive, from the mixed drones the queen mated with, no doubt.

Rings and Spots may help give additional information on the degree of hybridisation.

Drone body colour is used as some strains exhibit differences between male and female colouration; this character has more consistency as hybridisation decreases.

Two wing measurements are used to indicate the purity of strain against known data for different species.

1. Cubital Index. By measuring the ratio of two of the wing vein segments, we obtain measurements that are consistent for given races of bee; it does not matter what measurement method you use as the result is a ratio. I used pixel count.

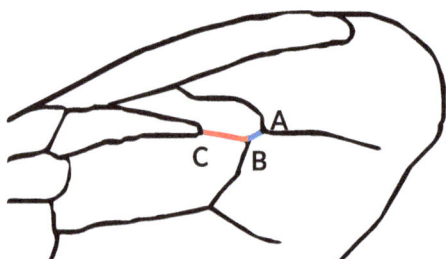

Fig 145. The points A, B & C should be judged to the centre of the vein junctions concerned. The distance "BC" is divided by the distance "AB".

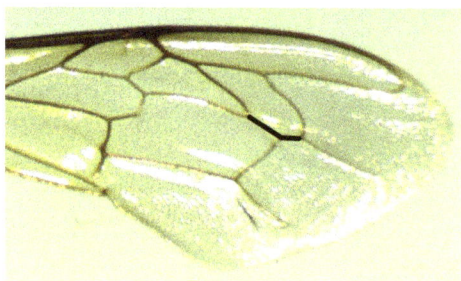

Fig 146. Discoidal Shift. This is used to identify Apis Mellifera Mellifera, which has a negative value, as most other types are zero or positive.

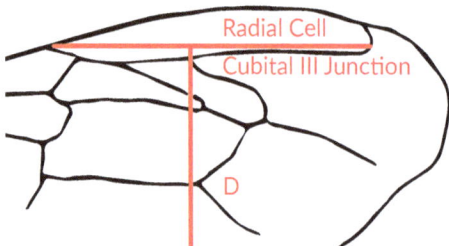

Fig 147.

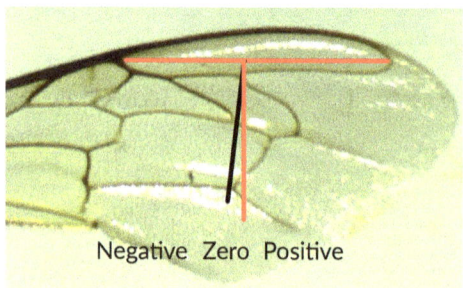

Negative Zero Positive

Fig 148.

Discoidal shift is an angular measurement in degrees, within the cells of the wing venation. A parallel line is drawn within the radial cell. Then a right-angled line is drawn through the Cubital III junction; from this, a third line can be drawn from the starting point through the Cubital vein and the angle noted. If it is to the left as shown above, then it is said to be negative, central zero, and to the right, positive.

Set out below is an abridged version of the characteristics of each strains, taken from D Cushman's site.

Heavily hybridised bees will show multiply results and not offer any positive results, except to say that are of mixed race.

Character	Apis mellifera mellifera	Apis mellifera ligustica	Apis mellifera carnica
General Appearance	Large, broad, short limbs	Medium size, slim, long limbs	Medium size, slim, long limbs
Worker body colour	Black		Black
Rings		1, 2 or 3 - yellow. Scutellum may be yellow	Maybe one leather coloured ring
Spots	None or small (2nd tergite)		May have small spots
Drone body colour	Dark	Amber/yellow	Dark
Rings or Spots		Yellow rings	Small spots
Cubital Index (worker) average	1.7	2.3	2.7
Cubital Index (worker) min	1.3	2.0	2.4
Cubital Index (worker) max	2.1	2.7	3.0
Discoidal Shift, worker	Negative	Positive	Positive
Worker hair colour	Few dark hairs	Yellowish	Grey
Drone hair colour	Brown/black	Yellowish	Grey or grey/brown

From the data collected, we can see the following results.

Nuc hive. Discoidal shift angle ranged from zero to 6 degrees. Average value = 2.9 degrees. Ten-registered zero, the rest were positive.

Cubital index ranged from 1.46 to 3.69 Average value = 2.39

Looking at the table, we can assume that we have mostly *Apis mellifera ligustica*, with a few mixes.

Swarm hive. Discoidal shift angle ranged from 2 to 9 degrees. Average value = 5.06 degrees. They were mainly positive.

Cubital index ranged from 0.9 to 1.93 Average value = 1.55

Looking at the table, we can assume that we have a lesser hybrid, more *Apis mellifera mellifera* mix.

As my hives are close together, I must assume that there must be some 'guests in each hive'; it has been estimated that up to 40% of bees in a hive have drifted there.

Are the swarm wild bees or just escapees from another hive? See the final article for the conclusion, were we will look at the *abdomens* to note the hair pattern and hair length.

This second article explains the methodology used to check on the race of bees present in a hive, by means of measuring the over hairs and the Tomentum width, (these are fine matted hairs) which together with the wing measurements help identify your strain of bees. Other methods are available.

The *abdomen* of 30 bees from each hive was separated from the rest of the body and examined under the dissecting microscope. You could also use x 5 optical eye-glass (at the end of the exercise, save the 30 *abdomens* and mash them up to check for *Nosema*).

The *tomenta* are considered as narrow (less than 50% of *tergite*), Medium at about 50% and broad if more than 50%.

The 4th *tergite* is the segment used to judge the *Tomentum* width.

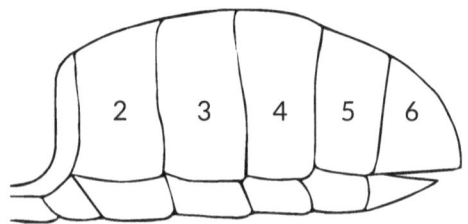

Fig 149. The abdominal tergites are numbered from 1 to 6.

The over hair length is judged on the fifth *tergite* and is compared with respect to a 0.40 mm wide wire.

Apis mellifera mellifera has narrow *tomenta* and long over hair. This compares with Italian strains that have broad bands of very short hair.

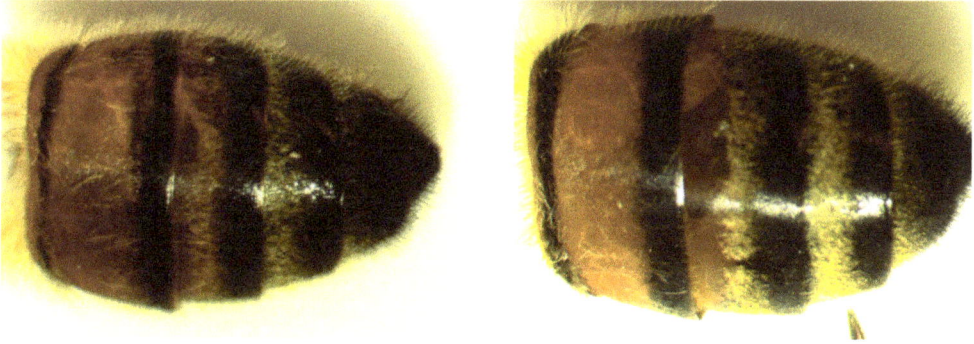

4th Tergite, 2 different bees. Fig 150. Left picture has a narrow band; Fig 151. Right picture has a broad band.

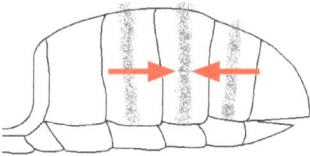

Fig 152. These would be considered narrow tomenta.

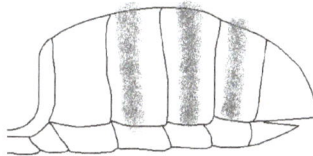

Fig 153. These would be considered medium tomenta.

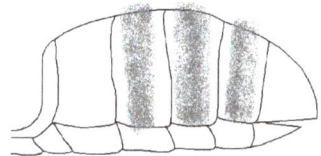

Fig 154. These would be considered broad tomenta.

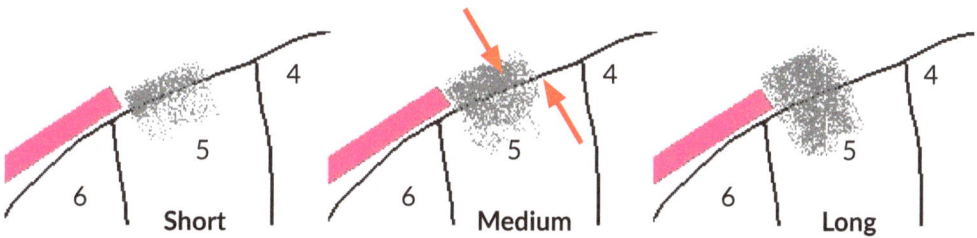

Fig 155. Short over hair, comprises of anything up to 0.35 mm. Medium over hair fall in the range of 0.35 mm - 0.40 mm. long over hair, anything longer than 0.40 mm.

Dave Cushman's data chart.

Character	Apis mellifera mellifera	Apis mellifera ligustica	Apis mellifera carnica
5th Tergite Over hairs (mm)	0.4-0.6	0.2-0.3	0.25-0.35
Tomentum Width (4th tergite)	narrow, less than 1/2 of tergite	broad, more than 1/2 of tergite	broad, much hair

Therefore, to compare the two colonies using Dave Cushman's chart again we have the following results.

Nuc colony *tomenta* over hair length average were 0.3 and a broad covering of hair.

Swarm colony *tomenta* over hair length average were 0.4 and a narrow covering of hair

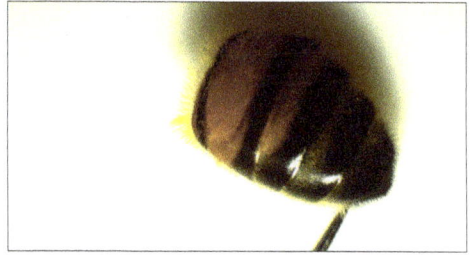

Fig 156. This bee has a 0.3mm needle placed nearby for direct comparison, resulting in long over hairs.

The conclusion using both sets of data is as follows. The Nuc hive is a mix but shows stronger tendency to be more *Apis mellifera ligustica*, with a few mixes.

The Swarm hive shows strong evidence for being more of an *Apis mellifera mellifera* mix.

This article has been put together from several sources, mainly the excellent Dave Cushman's web site. Please visit to find more information: www.dave-cushman.net.

Martyrs to the cause

The drone and worker bees have different roles to play within the hive; as its name denotes it is the worker who is the industrious one; the other is built for reproduction - passing his and the queen's *genes* onto the next generation. They both share an altruistic character; they will sacrifice themselves for the greater good of the colony. The worker will defend the hive and by stinging its foe, it cannot withdraw the stinging lance due to the design - a toothed barb on the ends, which help grip one side whilst the other lance is pushed in further. Because of this, the stinging mechanism as a whole is torn out and the worker dies. The remaining sting body releases a *pheromone* to attract other workers to help defend the colony. The drone on the other hand also dies after copulation with a virgin queen as his *endophallus* is left behind in the queen's sting chamber as a mating sign for other drones. The drone's *endophallus* has two orange coloured *cornua* (horns) which other drones can see in ultra vio-

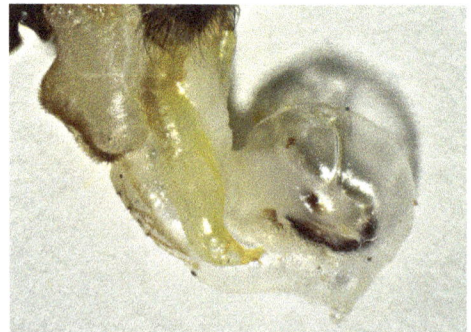

Fig 157. The everted endophallus showing the two orange horns (cornua), the bulbous end would be inside the queen's sting chamber. X 30 magnification.

let. This acts as an attraction for other drones. The drone has to die because all his *sperm* and all his brothers in the hive are the same genetically. It is in the interest of the hive for the queen to mate with differing drones, as each will have a different genetic makeup allowing the hive to adapt to any evolutionary changes in the future.

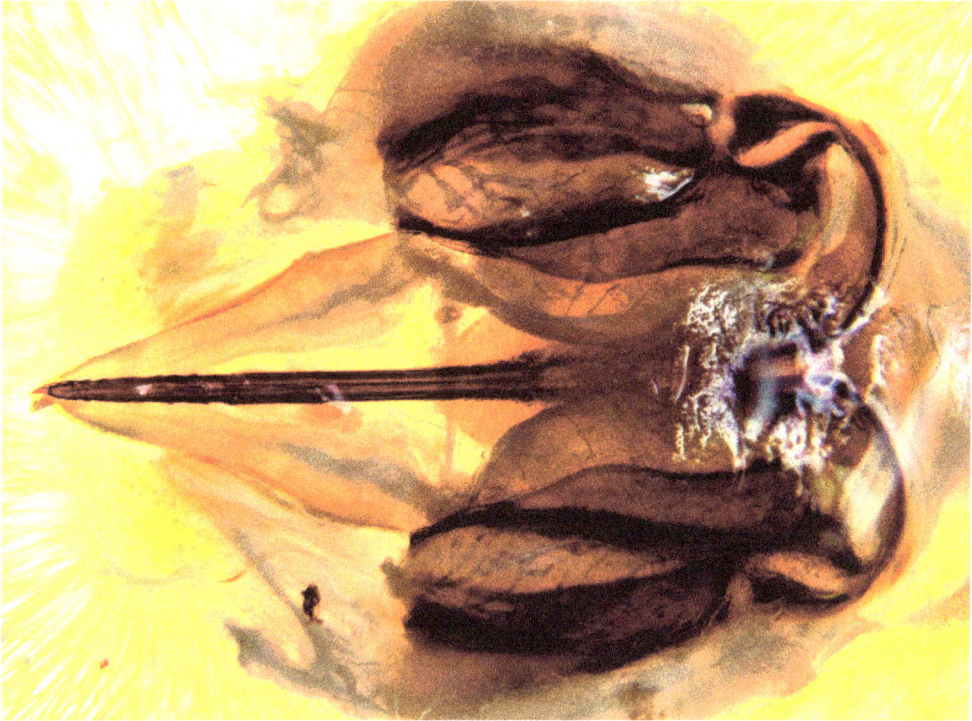

Fig 158. The remaining sting mechanism, taken from my bee suit. X 40 magnification.

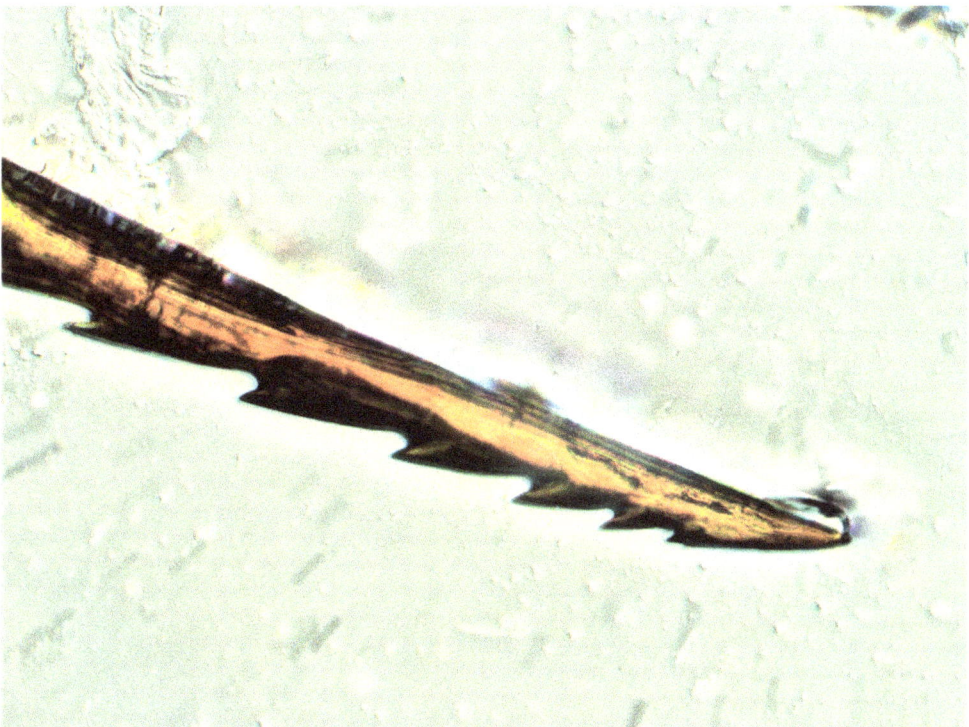

Fig 159. The barbs on the lances X 200 magnification.

The bee's knees

Do bees have knees? The answer is quite possibly. In fact, they could have six.

In humans, the knee is the joint between the femur and the tibia. Since bees have a femur and a tibia in each leg, they could have knees. However, unlike us, their joints don't have a kneecap (patella).

Bees have legs with many joints like any insect; however, by NOT having knee caps, this technically means that they do not possess any knees.

According to the Oxford English Dictionary, 'the bee's knees' phrase originated in the late eighteenth century meaning something very small. However, its current meaning is believed to stem from American slang during 1920. It is an informal expression referring to something outstanding or truly excellent. The origin might possibly have come from the fact that both words rhyme together.

Some specialised features of importance to the bee are shown below.

Fig 160. The antenna *cleaner in opened position on the fore leg, showing the notches in the comb with the* fibula *sticking up; this closes when the leg is bent to form a hole, allowing the antenna to be dragged through. The other hairs act as a brush to help clean the head of dust and pollen. Image X 100 magnification.*

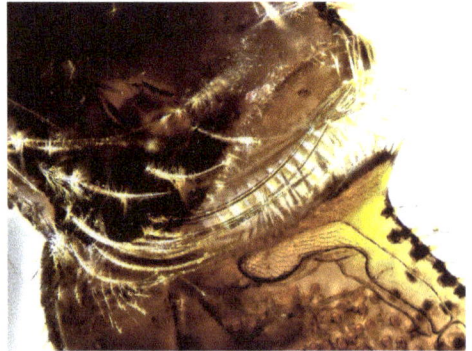

Fig 161. The pollen press on the hind leg *showing the* ratellum - *large stiff like hairs on the top; these are used as a raking action on the opposite leg, collecting pollen from the hairs. The small hairs on the bottom are used to retain the pollen. Inside bottom are the teeth on the auricle. The action of raking allows the pollen to be collected in the auricle and mixed with saliva; the leg is then bent, squashing the pollen together and out onto the outside of the tibia, to be caught by the special curved hairs know as the corbicula. Image X 100 magnification.*

Fig 162. *The single spine on the middle leg possibly used to collect pollen from the thorax; there is some conjecture over its usage or even if it is a left over vestige. X 100 magnification.*

Fig 163. *Pollen pellet stored on outer leg, having been pressed. Pollen grains collected on the inside leg, the bastitarsus; these have pollen brushes in rows which are clearly visible. X 40 magnification.*

A Hive Portcullis, natural gatekeeping

I recently moved my four hives to a wooded, open setting, facing south towards the sea on the Devon pebble bed heath land. After a few days, I visited just to check that they had settled in and no woodpeckers had been attracted to the hives as both species of woodpecker reside in the woods. Two weeks later in mid November, I went to check them again. You will see the old seed trays at the bottom of the entrances, weighed down with a large pebble from the surrounding beds. Together with the floor slider trays, I can monitor the hives without opening them up each time throughout the year.

As the mid afternoon temperature was about eleven degrees Centigrade, I was surprised to see the bees returning with pollen loads, yet to be identified. Ivy and gorse are still in flower locally.

Fig 164. The edge of the woodland glade.

I emptied the seed trays at each visit and was surprised to see one of the polystyrene hives have the entrance blocked up with propolis, with the exception of three bee-sized holes at one end.

It would seem that they have made a portcullis at the entrance to keep out intruders.

The dictionary definition is a strong, heavy grating sliding up and down in vertical grooves, lowered to block a gateway to a fortress or town.

Probably, like most beekeepers' this year, I was plagued in late August by wasps. Luckily they chose to eat the pears behind the apiary first and generally ignored the hives. I reduced the hive entrances down to 50 - 75mm long to make it easier for them to defend. The wooden hive was attacked in particular, once the pears had been harvested, being sited the closest to the fruit trees. It would appear that the bees have chosen their own method of defence, leaving only three holes available to guard. I am thinking of putting a sign up next to the hive - BEES AWARE of intruders!

The end hive in the first picture is a more natural hive which mimics a tree in insulation, this has a 60 mm long tube leading into the hive, with a circular disc to control the size, if need be, or to close the hive off for moving.

It will be interesting next season to see how they use this hive; more so as apparently the wasps and hornets do not like to enter tunnels. The tube is 40 mm in diameter so this might be too large to be effective. This hive had a swarm in and they did not seem to be vistied by wasps this season. Time will tell.

Fig 165. Left: The entrance nibbled out on the left.
Fig 166. Right: Propolis sealed entrance showing two of the entrance holes on the right.

References

Beekeeping web sites
Dave Cushman website: *www.dave-cushman.net*.
By kind permission of the current owner, Roger Patterson.

Wing clipping, some thought
The Cambridge Declaration on Consciousness.

May pollen
The Pollen Grain Drawings of Dorothy Hodges. Taken from the *Pollen Loads of the Honeybee*. Publisher IBRA SBN-13: 978-0860982623 by kind permission.
Pollen Identification for Beekeepers. 1st Edition by Sawyer, ISBN: 9780906449295

Magnetic sense in honeybees
Liang, C., Chuang, C., Jiang, J. et al. *Magnetic Sensing through the Abdomen of the Honeybee*. Sci Rep 6, 23657 (2016). https://doi.org/10.1

Natural thoughts
Summaries of a paper written by Peter Neumann from Basle University, who has kindly agreed to its publication. https://doi.org/10.1111/eva.12448

An alternative slant on honeybee history
Josephine: Desire, Ambition, Napoleon. Author Kate Williams. ISBN 978 0 099 55142 3
Publisher Arrow.

Honey magic
Traditional and Modern Uses of Natural Honey in Human Diseases: A Review. Tahereh Eteraf-Oskouei. Moslem Najafi. Biotechnology Research Centre, Tabriz University of Medical Sciences, Tabriz, Iran. Department of Pharmacology and Toxicology, Faculty of Pharmacy, Tabriz University of Medical Sciences, Tabriz.
Doctor Peter Molan, dedicated a lot of time researching the health benefits of Manuka honey, he has generously published his research documents free for all to read. *http://waikato.academia.edu/PeterMolan*
There is a list of contents and many detailed papers, which are printed on separate subject matters. Well worth seeking out if, there are many specific areas of treatments, not mentioned in article.

Entomology, etymology, the honeybee and words
A Honeybee Heart Has Five Openings, by Helen Jukes with kind permission.

Right buzz
Source: Extremely high frequency sensitivity in a 'simple' ear. Hannah M. Moir, Joseph C. Jackson and James F. C. Windmill. *https://doi.org/10.1098/rsbl.2013.0241*

How not to get lost
Incredible Journeys: Exploring the Wonders of How Animals Find Their Way, by David Barrie. Published by Hodder & Stoughton. Parts quoted with kind permission.

Which side of the fence do we sit?
The full open access article by CA Hallmann can be found here: *www.journals.plos.org/plosone/article?id=10.1371/journal.pone.0185809*

Number 30, the bee's special number?
Beekeeping Study Notes - Microscopy certificate by the Yates team, see Appendix 1
Dave Cushman's beekeeping web site: *www.dave-cushman.net*

Who will be the bees advocate?
The Lives of Bees, Prof T Seeley ISBN-10: 0691166765 with kind permission.

Record Breaking
Extremely high frequency sensitivity in a 'simple' ear. Hannah M. Moir, Joseph C. Jackson and James F. C. Windmill. *https://doi.org/10.1098/rsbl.2013.0241*

An opportunity to see in ultraviolet!
University of Bristol Biological Sciences Department
Wikipedia.

At sleep or just resting?
This article has been summarised from the study by the following authors. Barrett A. Klein1, Kathryn M. Olzsowy, Arno Klein, Katharine M. Saunders and Thomas D. Seeley 'Caste-dependent sleep of worker honeybees'. The study has also put four short videos on the web showing the stages of sleeping workers.

The Hollyhock's spiky secret.
Plants for bees. Kirk and Howes. IBRA
Klaus Lunau, Vanessa Piorek, Oliver Krohn, Ettore Pacini. Just spines-mechanical defence of malvaceous pollen against collection by corbiculate bees. Apidologie, Springer Verlag, 2015, 46 (2), pp.144-149. ff10.1007/s13592-014-0310-5ff. ffhal-01284432f

Polling results in!
This information was taken from the excellent book Plants for Bees. A Guide to Plants that Benefit the Bees of the British Isles. By kind permission of IBRA. The second edition also includes pollen slides for some plants.

Honey and Pollen Coefficients
http://entnemdept.ufl.edu/honeybee/extension/Honey%20Show%20and%20Judging/Bryant%20%20Jones%20%282001%29%20The%20R-Values%20of%20Honey%20-%20Coefficients.pdf

What the Microscopist DIDN'T want to see!
FERA handbook.
Baily & Ball. Honeybee Pathology.

Virus and the honeybee
National bee unit
Diversity and Global Distribution of Viruses of the Western Honey Bee, Apis mellifera ©2020 by the authors. Licensee MDPI, Basel, Switzerland. This article is an open access article distributed under the terms and conditions of the Creative Commons Attribution (CC BY) license (http://creativecommons.org/licenses/by/4.0/).

Honey bee genetics: The drone's role.
Professor J Woyke publications and private email.

General reference books
Anatomy and Dissection of the Honeybee. Dade. IBRA. 2009. ISBN 0-86098-214-9 by kind permission of IBRA.
Anatomy of the Honey Bee. Snodgrass. Cornell University Press. 1984 IBSN 978-0-8014-9302-7

Form and function of the honeybee by Lesley Goodman. Publisher IBRA ISBN-13: 978-0860982432 by kind permission.
The Buzz about Bees. Tautz. Springer. 2008. ISBN 978-3-540-78727-3 by kind permission of Professor Tautz
Understanding Bee Anatomy, a full colour guide. Stell. The Catford Press. 2012. ISBN 978-0-9574228-0-3.

All photographs by Graham Kingham unless otherwise stated.

Acknowledgements

A debt of gratitude must go to Professors Woyke, Tautz and Seeley for kindly allowing me to use their articles and research data.

IBRA has the copyright for Dade's Anatomy & Dissection of the Honeybee. They also hold the copyright to Bee World articles. Permission to use both sets of data has been kindly granted and acknowledged.

My external thanks go to all the bee scientists who have freely published their research. The many beekeepers who have educated me in a multitude of ways, especially Chris Utting, Julia Elkin, Lilah Killock and Alan White for all those bee thoughts over a beer or two. My wife Catherine Kingham the queen in my life.

I am in constant debt to Pat and David Woodward who are both highly qualified microscopists. They have helped and advised me over the years and have always encouraged me to pursue my interests, once again, a very big thank you.

Technical bits

For those of a technical mind the microscopes used were:

- Meiji compound biological microscope MX4200H

- Brunel zoom dissecting trinocular stereomicroscope BMDZ.

Brunel has a vast range of quality microscopes new and used suitable for every pocket. They offer outstanding lifetime service and advice on all matters concerning the microscope world. www.brunelmicroscopes.co.uk

Cameras used:

- Toupcam E3CMOS. 12 mega pixels. USB 3. Software: ToupView.

- Olympus OM D E-M1 camera with 60mm macro lens.

Other books by the author:

- Honey Bee Drones: Specialists in the Field (2019)
- Honey Bee Anatomy: Brought to Life (2021)

www.ingramcontent.com/pod-product-compliance
Lightning Source LLC
Chambersburg PA
CBHW051658210326
41518CB00021B/2593